Y

Accessory to War: The Unspoken Alliance Between Astrophysics and the Military (2018)
With Avis Lang

Astrophysics for People in a Hurry (2017)

StarTalk: The Book (2016)
With Charles Liu and Jeffrey Lee Simons

Welcome to the Universe: An Astrophysical Tour (2016)
With Michael A. Strauss and J. Richard Gott III

The Inexplicable Universe (2012)
A six-part video lecture series

Space Chronicles: Facing the Ultimate Frontier (2012)
Edited by Avis Lang

The Pluto Files: The Rise and Fall of America's Favorite Planet (2009)

Death by Black Hole: And Other Cosmic Quandaries (2007)

Origins: Fourteen Billion Years of Cosmic Evolution (2004)
With Donald Goldsmith

Letters *from an* Astrophysicist

Neil deGrasse Tyson

W. W. NORTON & COMPANY
Independent Publishers Since 1923

Copyright © 2019 by Neil deGrasse Tyson

Epigraph: 1704 March/April, *Philosophical Transactions of the Royal Society of London*, A Letter to Dr. Edward Tyson from William Cowper, page 1586. Printed for S. Smith and B. Walford, Printers to the Royal Society, London.

Photo of Comet Schwassmann-Wachmann 3 on page 71: NASA/JPL-Caltech, W. Reach (SSC/Caltech).

Photo of the Helix Nebula on page 158: NASA, WIYN, NOAO, ESA, Hubble Helix Nebula Team, M. Meixner (STScI), and T. A. Rector (NRAO).

All rights reserved
Printed in the United States of America
First Edition

For information about permission to reproduce selections from this book, write to Permissions, W. W. Norton & Company, Inc., 500 Fifth Avenue, New York, NY 10110

For information about special discounts for bulk purchases, please contact W. W. Norton Special Sales at specialsales@wwnorton.com or 800-233-4830

Manufacturing by Lake Book Manufacturing
Book design by Lovedog Studio
Prepress production by Joe Lops
Production manager: Julia Druuskin

ISBN 978-1-324-00331-1

W. W. Norton & Company, Inc.
500 Fifth Avenue, New York, N.Y. 10110
www.wwnorton.com

W. W. Norton & Company Ltd.
15 Carlisle Street, London W1D 3BS

1 2 3 4 5 6 7 8 9 0

To my mother, who first taught me how to write with meaning and impact. And to my father, whose life experience navigating people, places, and things conferred upon me the necessary wisdom to navigate a life of my own.

If in this I have been tedious, it may be some excuse,
I had not time to make it shorter.

—William Cowper

Contents

III. Pathos

A plaintive appeal to emotions that already reside within us.

IV. Kairos

A propitious moment for decision or action.

Preface

Now that people communicate with one another primarily by social media, letter-writing has become a lost art. The greatest casualty may be our growing inability to find words that precisely communicate our feelings and emotions. Why else the need for that burgeoning catalog of emoticons to supplement our written correspondence? A smiley face. A snarky face. A heart. A thumbs-up. But when the world stimulates your curiosity; when your state of not knowing leaves you restless; when your existential angst overflows; sometimes you just have to write a full-up letter to somebody.

Contained herein is a sampling of my correspondence, almost all with complete strangers, spanning more than two decades, with most letters selected from a ten-year period when my email address was publicly accessible.* Over that time, most inquiries contained straightforward questions of science.

* When a letter was received via means other than email (e.g., US "snail" mail or social media), that medium is indicated.

Those were fielded by expert staff at New York City's Hayden Planetarium, where I serve as director. Other letters, mostly of a personal nature, including those with specific reference to a speech I had given, a book I have written, or a video in which I appeared, form the corpus of letters from which my responses are drawn.

The letters to me that communicate significant emotion, curiosity, or anxiety are reproduced in full.* Other letters, of the rambling type, I summarize for brevity in a single paragraph. Some letters are written by people who are angry with the world or with something I have said or done. Others explore ideas and beliefs. Still others are sad, sensitive, and poignant. And in many cases there's a longing we've all experienced at one time or another: the search for meaning in our lives; an evergreen urge to understand one's place in this world and in this universe.

Added, too, are letters I've written, not to anybody but to everybody. These include letters to the editor, mostly of the *New York Times*, as well as open letters posted to my Facebook page and other public places on the internet. One of the earliest dates back to my overlong September 12, 2001, letter to family and colleagues, twenty-four hours after I bore witness, from a distance of four blocks, to the attack and collapse of the World Trade Center's Twin Towers.

* When sensible, letters are lightly edited for spelling and grammar. Long letters are also edited for clarity and length. Meanwhile, zealous punctuation, when issued forth from EMOTION, is left largely intact!!!

Above all else, *Letters from an Astrophysicist* is a vignette of the wisdom I have mustered to teach, enlighten, and ultimately commiserate with the curious mind. It's the world as viewed through the lens of an astrophysicist-educator. A world now shared with you.

Letters from an Astrophysicist

Prologue

A Memoir, of Sorts

Happy 60th Birthday, NASA

Monday, October 1, 2018
Facebook Note

Dear NASA,
Happy birthday! Perhaps you didn't know, but we're the same age. In the first week of October 1958, you were born of the National Aeronautics and Space Act as a civilian space agency, while I was born of my mother in the East Bronx. So the yearlong celebration of our shared sixtieth anniversary provides me a unique occasion to reflect on our past, present and future.

I was three years old when John Glenn first orbited Earth. I was seven when you lost astronauts Grissom, Chaffee, and White in that tragic fire of their Apollo 1 capsule on the launch pad. I was ten when you landed Armstrong and Aldrin on the Moon. And I was fourteen when you stopped going to the Moon altogether. Over that time I was excited for you and for America. But the vicarious thrill of the journey, so prevalent in the hearts and minds of others, was absent from my emotions. I was obviously too young to be an astronaut. But I also knew that my skin color was much too dark for you to picture me as part of this epic adventure. Not only that, even though you are a civilian agency, your most celebrated astronauts were military pilots, at a time when war was becoming less and less popular.

During the 1960s, the Civil Rights movement was more real to me than it surely was to you. In fact it took a directive from Vice President Johnson in 1963 to force you to hire black engineers at your prestigious Marshall Space Flight Center in Huntsville, Alabama. I found the correspondence in your archives. Do you remember? James Webb, then head of NASA, wrote to German rocket pioneer Wernher von Braun, who headed the Center and who was the chief engineer of the entire manned space program. The letter boldly and bluntly directs von Braun to address the "lack of equal employment opportunity for Negroes" in the region, and to collaborate with the area colleges Alabama A&M and Tuskegee Institute to identify, train, and recruit qualified Negro engineers into the NASA Huntsville family.

In 1964, you and I had not yet turned six when I saw picketers outside the newly built apartment complex of our choice, in the Riverdale section of the Bronx. They were protesting to prevent Negro families, mine included, from moving there. I'm glad their efforts failed. These buildings were called, perhaps prophetically, the "Skyview Apartments," on whose roof, 22 stories over the Bronx, I would later train my telescope on the universe.

My father was active in the Civil Rights movement, working under New York City's Mayor Lindsay to create job opportunities for youth in the ghetto—as the "inner city" was called back then. Year after year, the forces operating against this effort were huge: poor schools, bad teachers, meager resources, abject racism, and assassinated leaders. So while you were celebrating your monthly advances in space exploration from Mercury to Gemini to Apollo, I was watching America do all it could to marginalize who I was and what I wanted to become in life.

I looked to you for guidance, for a vision statement that I could adopt that would fuel my ambitions. But you weren't there for me. Of course, I shouldn't blame you for society's woes. Your conduct was a symptom of America's habits not a cause. I knew this. But you should nonetheless know that among my colleagues, I am the rare few in my generation who became an astrophysicist in spite of your achievements in space rather than because of them. For my inspiration, I instead turned to libraries, remaindered books on the cosmos from bookstores, my rooftop telescope, and the Hayden Planetarium. After some fits and starts through my years in school, where my ambitions seemed at times to be the path of most resistance through an unwelcoming society, I became a professional scientist. I became an astrophysicist.

Over the decades that followed you've come a long way. Whoever does not yet recognize the value of this adventure to our Nation's future, soon will, as the rest of the developed and developing world passes us by in every measure of technological and economic strength. Not only that, these days you look much more like America—from your senior-level managers to your most decorated astronauts. Congratulations. You now belong to the entire citizenry. Examples of this abound, but I especially remember when the public took ownership of the Hubble Telescope, your most beloved unmanned mission. They all spoke loudly back in 2004, ultimately reversing the threat that the Telescope might not be serviced a fourth time, extending its life for another decade. Hubble's transcendent images of the cosmos had spoken to us all, as did the personal profiles of the Space Shuttle astronauts who deployed and serviced the telescope, and the scientists who benefited from its data stream.

Not only that, I had even joined the ranks of your most trusted, as I served dutifully on your prestigious Advisory Council. I came to recognize that when you're at your best, nothing in this world can inspire the dreams of a Nation the way you can—dreams fueled by a pipeline of ambitious students, eager to become scientists, engineers, and technologists in the service of the greatest quest there ever was. You have come to represent a fundamental part of America's identity, not only to itself but to the world.

So as we both turn sixty, and begin our sixty-first trip around the Sun, I want you to know that I feel your pains and share your joys. And I look forward to seeing you back on the Moon. But don't stop there. Mars beckons, as do destinations beyond.

Birthday buddy, even if I have not always been, I am now and evermore your humble servant.

Neil deGrasse Tyson
New York City

I.

Ethos

The characteristic spirit of a culture,
manifested in its beliefs and aspirations.

Chapter 1

Hope

It's all you have when you realize you are not entirely in control of outcomes. But without it, how else do we cope with the challenges of life?

Coma

Sunday, February 25, 2007
Dear Mr. Tyson,
I have long suspected that we do live in a universe that wants to kill us, so I am not surprised you say this in your talks, but where is the hope, or isn't there any?

I spent 13 days in a coma in 2001 and miraculously returned to life to continue with my dear husband. He sang me a love song and invited me to return, and I opened my eyes and smiled at him. However, I am forever changed by the extent of information I came back with from that sojourn, and much of it was not good. Is the "not good" part the majority of what's out there in your view? If so, how do you enjoy life, or don't you?

Highest regards,
Sheila Van Houten

Dear Ms. Van Houten,
I see two kinds of hope. One of them is a kind of religious hope where one prays or performs some kind of cultural ritual for things to get better.

But there is another kind of hope—it's the challenge of learning about the real world and using our intelligence to change things for the better. In this way, it's the individual who is empowered to bring hope to the world.

So yes, the universe wants to kill us. But on the other hand, we all want to live. So let's find a way together to deflect the asteroids, find the cure to the next lethal virus, mitigate hurricanes, tsunamis, volcanoes, etc. This can only be enabled by the efforts of a scientifically and technologically literate public.

Therein lies a hope on Earth far greater than ever promised by the act of prayer or introspection.

Sincerely,
Neil deGrasse Tyson

Fear

Sunday, July 5, 2009
Dear Mr. Tyson,
I just watched you on public television. I admire how far you have come in life. I have always tried to do what I could to help others. I am 38, a mother of three, and a full time student. I was born and raised in a small town of about 1500. When my marriage of 16 years fell apart, I decided to finish my Associate of Applied Science degree and apply at the University of Washington's School of Social work.

I am moving to Snohomish on August 1st, and have no job, but have been applying to everything I can daily. It hit home when you talked about ambition. I have three kids to feed and all I want to do is work and go to school. My passion is to work in human services and I have done respite care and worked with the elderly, but will work in fast food to get to where I need to be.

I worry all the time that I won't be able to provide for my children, and I am scared to death to move, but I won't let that stop me. I don't care if I have to re-apply to the UW every year until I am 70-years-old, I will attend and work my way to my master's degree. I just don't know how to get rid of this sick feeling in the pit of my stomach that I will move and fall flat on my face.

I have the drive and determination. I just need a break—not a free ride, just a job. I want nothing for free. I just want a job a chance to work my way through.

I don't know why I am writing you. I don't want anything, just someone to listen to my fears. I don't have anyone to tell them to, and maybe you could understand.

Thank you for your time in reading this.

Lisa Kalma

Dear Lisa,

The people who fail in life are those whose ambitions were insufficient to overcome all the forces that work against them. And yes, failure is common to us all. But ambitious people use their failures as lessons to heed, as they push forward toward their goals.

Don't fear change. Don't fear failure. The only

thing to fear is loss of ambition. But if you've got plenty of that, then you have nothing to fear at all.

Good luck on your journey, as I offer you the opening quote in my memoir, *The Sky Is Not the Limit**

Beyond the judgments of others
Rising high above the sky
Lies the power of ambition

All the best, on Earth and in the universe.
Neil

Losing My Religion

Wednesday, April 29, 2009
Dear Dr. Tyson,
I grew up on a cattle farm in the rural mountains of North Carolina—and at times I used to think I was damned or disabled because the belief in a higher power just didn't "click." I went to church, Sunday school, and was surrounded by religion in every direction . . . yet something in me just kept asking questions.

I remember having to lie about my beliefs and wanting to give up (sometimes in tears) thinking if I lied about it enough I could eventually believe. I was kicked out of Sunday school for "asking too many questions."

* Neil deGrasse Tyson, *The Sky Is Not the Limit: Adventures of an Urban Astrophysicist* (Amherst, NY: Prometheus Books, 2004).

But then I started to discover others like myself (although much more intelligent and educated). I just wanted to thank you—your words can have so much more of an impact than you could fathom. You (and others) give people, who are isolated by geography, hope that they can stand firm in their non-beliefs and continue to ask questions. I know you are a scientist and a teacher—but to some people, you are hope.

George Henry Whitesides

Dear Mr. Whitesides,

Thanks for sharing your personal story.

It was (and is) never my intent to convert anybody's belief system one way or another. My aim is simply to empower people to think for themselves, rather than to have others do the thinking for them. Therein blooms the "soul" of skepticism and the "spirit" of free inquiry.

I am happy to have nourished this growth within you.

As we say in the cosmos . . . Keep Looking Up.

Neil deGrasse Tyson

On Being Black

Marc saw the quality of my contributions as a good sign that times are changing, but he was sure that I had suffered, and continued to suffer, from racial bias and prejudice. He longed for the day when skin color became an irrelevant reference to a person's identity. On Christmas Day 2008, he asked about my life experience as an African-American scientist.

Dear Marc,

Thanks for your note.

I am happy to report that references to me as a "black" scientist are, today, extremely rare—enough to surprise me that you make any mention of it at all. Of course, if your direct life experience suggests it, then I cannot reason it away, but other metrics remain strong indicators of my contention.

But let's go back a few years. For example, in 2001, when I was appointed to a White House commission of 12 to study the future of the U.S. Aerospace Industry, some (especially critics of George W. Bush) were quick to say, "they needed a black person." But when you actually looked at the commission membership, I was the only academic and I was not the only black person—another was a 4-star Air Force General. So the critique evaporated on analysis.

In another case, in 1996, while attending an evening gala for my Museum* (I was largely unknown to the public at the time), a liberal-minded woman at my table saw that I worked for the Museum—but only high-ranking Museum administrators were in attendance, and so she was quick to presume that I was head of Community Affairs or some such title commonly reserved for token Blacks. I replied that I was an astrophysicist, Director of the Hayden Planetarium, and Project Scientist for the Rose Center for

* The American Museum of Natural History, New York City, where I have served as the Frederick P. Rose Director of the Hayden Planetarium since 1996.

Earth and Space, under construction, after which she had nothing to say for the rest of the dinner.

Those kinds of encounters were common then, but simply do not occur anymore, except, possibly, among older people whose life experience was shaped in a Black & White America, rather than in simply, America. Various high-profile biographical mentions of me in recent years make no mention of my skin color.*

So the trends don't support your contentions, or perhaps, it indicates that your experience does not represent prevailing trends and truths.

Thank you for your supportive comments, and while the struggle continues, the times are indeed a-changin'.

Neil deGrasse Tyson

On IQ

Just a few days later, Marc continued, wondering about the difference in IQ scores between blacks and whites. He debates it often between friends and family, and sought more talking points to help him argue against it.

* E.g., 2007 *Time* magazine's 100 Most Influential People in the World; 2008 *Discover* magazine's 10 Most Influential People in Science.

Dear Marc,
The issue is bigger than race vs. IQ. It's more likely about the meaning of IQ at all. Have a look at the book titled *Genius Revisited: High IQ Children Grown Up*, which studied what became of hundreds of graduates of Hunter College Elementary school in NYC, a selective public school where the students have an average IQ of 150+.

Tracking them into adulthood, one might imagine great achievements among them. Not so. There were no Nobel Laureates. No Pulitzer Prize winners. In fact, no person was singularly distinguished in his or her field. Meanwhile, they are all successful by any normal measure in American society—happily married, secure jobs, manager-level or higher, homeowners, etc. But one can't help reflecting on what distinguishes singularly successful people from others, because if IQ mattered on the level that IQ purveyors claim, then *all* the shakers and movers of society would be people drawn from this population. But the data show this not to be the case.

IQ correlates nicely with GPA in high school and college, but after your first job, nobody ever asks what your college GPA was. What matters are your people skills, leadership skills, real-world problem solving skills, integrity, business acumen, reliability, ambition, work ethic, kindness, compassion, etc. So for me, conversations about race and IQ are of no practical consequence, any more than are conversations about race and hair color, or race and food preferences.

I do not know my IQ. It's never been measured. I graduated 350-ish out of 700 in my high school class. So few teachers (or classmates for that matter) would have said of me, "He'll go far." Why? Because the educational system fixates on test scores. Meanwhile, for two years running, I am listed in the "Harvard 100," a compilation of the hundred most influential living graduates of Harvard University.

Good luck in your conversations with family. If any of them has a question, I will be happy to take a stab at it. But there are clearly more important issues out there to debate than IQ.

Neil deGrasse Tyson

100 mph

Thursday, May 3, 2012
How's it going, Ty? I feel I can call you that 'cause I feel like I know you already.

I've watched literally every second of all your YouTube videos. I would be at your talks, but my job requires me to travel a lot. My name is Jarrett Burgess and I play professional baseball. I'm emailing you because ever since I was four-years-old, I wanted to be an astronaut. You inspired me and gave me confidence in doing what I love to do, despite the public and family pressure on me to play baseball. I want to be known for discoveries and making a difference in science. I don't want baseball to define me.

Keep up with your videos—you're even reaching out to people like me. Yeah, I can throw a baseball 100 mph from the outfield, or run a 60-yard dash in 6.2 seconds, and hit a baseball over 410 feet. Even when I'm on the field, I'm thinking about science. I want to pursue my goal in science. I need help and a guide on what I should start off doing. I'm 21-years-old and the most dedicated person, with great integrity and, most importantly, an amazing imagination. And I love the cosmos.

Please help me, Neil, in any way you can. I will appreciate it.

Jarrett Burgess

Dear Jarrett,
Thanks for that all-out appeal to connect with the cosmos. You express a dilemma that afflicts many in society: Should you do what you're best at? Do what others expect of you? Or do what you love most?

I love baseball (a few dozen of my Tweets are on the subject), so I'd be hard-pressed to tell you to take your 100 mph arm and study the universe. But I also happen to love what I do. And because I love what I do, I am self-driven and incentivized to make myself better at it every day—without limit.

If I remember correctly, minor league players make hardly any money at all. So your time in the farm system is conceived to hone your skills in anticipation of being called up, rather than to accumulate wealth. It seems to me that you could've instead attended a good baseball college, where you can play competitively while simultaneously majoring in astrophysics. If memory serves, in the early 1980s, Roger Clemens pitched for the University of Texas at Austin, took them to the nationals, then entered the Major League.

Meanwhile, in the 1980s, Brian May had a successful career as lead guitarist for the legendary rock group Queen, then—then—then—decided to get a Ph.D. in Astrophysics. Earned just a few years ago.

I'd bet most people who are encouraging you to stay in baseball carry high expectations that you will make tons of money. But that means your career would be driven by the search for wealth, rather than the search for cosmic fulfillment. In my experience, when money is the sole carrot, people can lose sight of life's deeper sources of happiness.

Until you major in physics or astrophysics in college (taking all the attendant math courses) you will not know for sure what you're better at—academics or sports. That will be useful to know. If you're better at sports than academics, but still love the universe, then return to professional baseball, play for 10 years while getting your master's degree over the winter months, then, like Brian May, get your Ph.D. after you've made tons of money.

If you delayed professional baseball, and went to college to major in Physics (while still playing baseball) that would make headlines—especially in today's science starved culture. And if it doesn't, I'll make sure it does.

In any case, I am delighted to learn that I have helped, in whatever small way, to sustain the rage of your cosmic flame.

Best to you.
Neil deGrasse Tyson

If I Were President

During a particularly obstinate run of congressional cacophony, the "Sunday Review" section of the New York Times *solicited responses from non-politicians to the phrase: "If I Were President . . ." What follows is the unedited version of my published answer.*

Sunday, August 21, 2011
New York Times

The question, "If I were President I'd . . ." implies that if you swap out one leader, put in another, then all will be well with America—as though our leaders are the cause of all ailments.

That must be why we've created a tradition of rampant attacks on our politicians. Are they too conservative for you? Too liberal? Too religious? Too atheist? Too gay? Too anti-gay? Too rich? Too dumb? Too smart? Too ethnic? Too philanderous? Curious behavior, given that we elect 88% of Congress every two years.

A second tradition-in-progress is the expectation that everyone else in our culturally pluralistic land should hold exactly your own outlook, on all issues.

When you're scientifically literate, the world looks different to you. It's a particular way of questioning what you see and hear. When empowered by this state of mind, objective realities matter. These are the truths of the world that exist outside of whatever your belief system tells you.

One objective reality is that our government doesn't work, not because we have dysfunctional politicians, but because we have dysfunctional voters. As a scientist and educator, my goal, then, is not to become President and lead a dysfunctional electorate, but to enlighten the electorate so they might choose the right leaders in the first place.

Neil deGrasse Tyson
New York City

Chapter 2

Extraordinary Claims

Curious about UFOs, cryptozoology, astrology, extrasensory perception? It's all here. The Carl Sagan dictum "extraordinary claims require extraordinary evidence" remains a potent guide when investigating the natural world for its underlying order. But it comes with a recurring risk: knowing enough about a subject to think you are right, but not enough about the subject to know you are wrong.

ET Phone Home

Saturday, March 8, 2009
Neil, if the ETs are out there, why don't we send someone to the moon and Mars and get their input on who they are and why they are coming to Earth?

Mel

Dear Mel,
Until someone drags an alien carcass into a public lab or ET lands on the White House lawn, or on the roof

of the *New York Times* building, one cannot justify the trillion-dollar expense of travelling to Mars to greet them because the weight of the evidence is incommensurate with the strength of the claims.

Neil

Alien Aliens

Sunday, November 8, 2009
Dear Neil
I have been patiently waiting for my wonderful scientists to "prove" that Aliens exist. and I believe it is coming, however slowly. And I might be thinking way out of the box here—but instead of always looking for something that mirrors us, why not look for something that does not?

Melodie Lander

Dear Melodie
The number of ways that life might be alive, for which we have no foundation to select, vastly exceeds the one way we know life can be alive. So when designing an experiment on a limited budget, you always start with what you know.

We know life is possible with carbon-based molecules (evidence = us). We further know carbon is highly abundant in the universe and is the most chemically fertile element on the periodic table. So we begin there.

Neil

UFO Sightings

Trenton Jordan noted that he was losing his skepticism regarding UFOs. The cause? Freshly released video footage from shuttle missions, in which unexplained objects were flitting about outside the windows. He was aware of space debris and other possible explanations, but he grew convinced that NASA must be withholding information about aliens that the public deserves to know. He wrote to me in July 2008, in search of arguments that might quell his skepticism.

Dear Mr. Jordan,
Thanks for your kind words about my life's work. They are warmly received.

Regarding your evaporating skepticism of visiting aliens: When you see shapes or lights that fly through the air or through space, and you do not know what they are, they become a UFO—emphasis on the "U." Such sightings split into four broad categories:

1. The observer is crazy or otherwise delusional.

2. The observer sees and reports inaccurately, confounding an account that would be a simple description of natural phenomena.

3. The observer sees and reports awccurately, but is insufficiently familiar with natural phenomena to be mystified by what he or she sees.

4. The observer sees and reports accurately something that defies any normal or conventional explanation—constituting a genuine mystery.

Note that eyewitness testimony is, by far, the weakest form of evidence that a person can present in support of a claim. In spite of its high value in the court of law, in the "court" of science, eyewitness testimony is essentially useless. Psychologists have known for quite some time how ineffective the human senses are as data taking devices. Note that the pedigree of the observer is irrelevant here—as long as he or she is human, the fallibility of observation is manifest.

Note further that claims of a "cover-up" or "conspiracy" is the battle cry of people who want to believe, in the face of insufficient data to fully support their claims.

Another well-known shortcoming of the human mind is what psychologists and philosophers call "argument from ignorance." The NASA cases you describe come closest to category (4) above, since we have video of strange phenomena—video that we take to be generally reliable, reminding us again of what the "U" in UFO stands for. Once you confess to not knowing what you are looking at, no logical line of reasoning allows you to then declare that you know what you are looking at. And that includes assertions that the flying shapes "must be" intelligent, technologically advanced aliens from distant planets secretly observing the behavior of Earthlings. You simply bear insufficient evidence to make that jump, however tempting it may be.

A similar argument from ignorance comes from the Big Bang. When I am asked what was around before the Big Bang, I say, "We do not yet know." Often the reply is, "It must be something—it was surely God." To go from "We don't know" to "It must be God" is another example of an argument from ignorance. This kind of disconnect has no place in rational investigations, yet it perennially permeates the thoughts and statements of people who already know what they want to believe.

So if the flying mysteries actually turn out to be intelligent aliens, it will not have been demonstrated by any observation yet brought forth. What's required to draw the conclusions you seek is much better evidence of the kind that would survive the "court" of science: Aliens visiting multiple media centers, for example, demonstrating their technology on national television; joining the President and First Lady for a state dinner or high tea in the Rose Garden; allowing themselves to be CAT scanned at the Johns Hopkins Medical Center so that we can learn about their physiology; submitting some of their communication devices or other hardware to our most respected research laboratories. The day real evidence comes forth, you will not need congressional hearings parading high-ranking eyewitnesses on the subject.

Until any of that happens, UFO sightings of category (4) are simply intriguing unidentified lights and shapes in the sky—perhaps worthy of further study like any mystery in science—but without conspiracy theorists invoking cover-ups to bridge all gaps in the data, convincing themselves of what they are already sure is true.

Should NASA direct funds to study these mysterious reflective objects visible from the window of spacecraft? Would be nice one day to have a radar device that constantly monitors and photographs anything of any size that approaches the craft. But there's so much happening outside the window of a spaceship—dislodged tools, loose paint chips, particulate fuel exhaust floating by. Not to mention rapid, ever-changing lighting conditions.

In summary, if you want public money to investigate UFOs for the possibility that they may be alien visitations, then you need much, much better evidence to justify the cause.

Thanks for your interest.
Neil deGrasse Tyson

A Glowing Pattern in the Sky

In March 2005, Dave Halliday of New Jersey wrote of facing northward at night as a teenager in the mid-1970s, when he witnessed what looked like a star encircled with orange dashes projecting outward. His guess at the time was that maybe he saw a planet bombarded by a meteor shower. Harboring this mysterious vision for three decades, he wondered if I could shed some light on what he saw.

Dear Mr. Halliday,
You have asked about a 1970s sightings of orange dashes. There is nothing more fickle and unreliable than eyewitness testimony, no matter the pedigree of the person who made the observation. This is why eyewitness testimony comprises the lowest form of evidence in science (curiously, unlike courts of law).

For example, I recently received an email from a retired engineer who said he saw a brilliant meteor streak across the Brooklyn sky at 8:15pm the night before. He wondered if I had heard of any other reports. Sounds like a high precision account. However, five other city-based reports placed a similarly described event between 7 and 7:30 pm. So unless two bright meteors crossed the sky that night, somebody has the wrong time. When confronted with this fact, the engineer notified me that his wife corrected his memory and in fact what he saw took place at 7:15, not 8:15. Notice that this exchange of information is happening 24 hours after the event. Not a decade. Not thirty years. Not a century. And you would think that reading the time would be the least likely source of error, since we do it every day.

With this story as preamble, I know of no cosmic phenomenon that would create the view that you describe. The closest thing I can think of must start with the question: Do you have long eyelashes? If they are wet and you look at a small bright source, the light will pass through droplets in your eyelashes before reaching your iris, and create a pattern of active wagon-wheel spokes. Try it. The best effect comes when rising up out of an outdoor pool.

Another possible source is trace-light illumination on the bottom of a blimp. In the dark the blimp itself is hardly visible, but lights beneath it, which are normally used for advertising, can take on interesting patterns, depending on what they are programmed to do. Pale orange was the color of the lights used back then.

Other than these two suggestions, I have no further explanations to account for what you saw, given that all you have is eyewitness testimony.

Thanks for sharing your story.

Sincerely,
Neil deGrasse Tyson

End of the World

In July 2009, fifteen-year-old Kale Joyce wrote with great concern about how many people believed that the world would end in 2012, citing the internet and pop media as purveyors of these fears. Though a non-believer in any of it, she nonetheless wanted my views on the predictions of Nostradamus and of all the mystery surrounding the end of the Mayan calendar.*

* Michel de Nostradamus (1503–1566), French physician widely believed to have the power of prophesy. Author of *Les Prophéties* (1555), containing 942 poetic passages that make predictions. John Hogue, ed. 1997, *Nostradamus: The Complete Prophesies* (London: Element Books, 1997).

Hello Kale,
The entire literature that surrounds 2012 is a hoax perpetrated by scientifically illiterate people who exploit the irrational, primal fears that lurk deep within us all.

The world will not end in 2012. Not because I am an authority and I say so. The world will not end because any sensible, scientifically literate person can assess the rampant absence of evidence and make up his or her own mind.

The galactic center, Sun, and Earth align every year on December 21. The Mayans were clueless about the laws of physics. Nostradamus was even more clueless about the laws of physics. And besides, he says nothing about the year 2012.

You are only 15, but every decade some group of people predicts the world will soon end. It has happened for the years 1973 (a comet), 1982 (a planetary alignment), 1991 (solar storms), 2000 (millennium madness), and now for 2012.

Want to live a long time? Worry about other things, like “Am I eating well,” “Do I get enough exercise,” and “Am I wearing my seatbelt?”

Sincerely,
Neil deGrasse Tyson

Time's Up

Sunday, September 6, 2009
Dear Dr. Tyson
We saw your video clip from the Los Angeles library regarding the Sun's alignment with the center of the Galaxy (that happens annually, which makes me feel much better). But can you explain as to why the Mayan calendar ends on that date. . . . as well as other older texts coming from the Chinese and Nostradamus. Do you think they have any validity? They say that the Mayan Calendar was even more precise than our own.

Iris Hale and son, Michael Hale

Dear Iris and Michael,
It's the end of what Mayan scholars call the "long count," begun on August 11, 3113 BC, the beginning of the universe according to Mayan calculations. At the end of the long count, they imagined the end of the universe.

With regard to the beginning of the universe, they were off by at least 13 billion years. So there's no reason to think they will get the end correct.

None of them mention the year 2012. And all the doomsday comet predictions via Nostradamus apply to the year 2000. Of course, none of those apocalyptic predictions came true. Furthermore, Nostradamus writes with such poetic imprecision that for practically any event after it happens you can go back and find a passage that sort of matches what occurred, and then declare that Nostradamus had special powers to see the future.

If you instead try to use his books' content to predict events with any accuracy before they happen, his imprecise quatrains fail badly—and do so consistently. This makes the writings of Nostradamus useless as a source of insight to the operations of the world.

Finally, the current Gregorian calendar used worldwide is accurate to a day in about 44,000 years. No other calendar comes close to this. So we're good for now.

Neil deGrasse Tyson

Defacing Mars

Friday, January 5, 2007
Dr. Tyson,
I am a big fan of yours, you are a rock star to me (kind of). I would love to get your perspective on things such as gravitics and Cydonia on Mars. If I could, I would go to Cydonia on Mars!

Also, I was reading a book by Linda Goodman and it is called Sun Signs. * *I understand the science aspect that you were representing, but this lady is on to something. Most*

* Linda Goodman, *Sun Signs* (New York: Bantam Books, 1985).

astrologers are indeed bogus. But we know how great the Egyptians were. They studied Astrology!

Thank you, Dr. Tyson
Stevie Debe

Dear Mr. Debe,

Gravitics: A delusion of its adherents, whose background in physics is meager, at best—leaving them to think they have discovered a new force of nature.

Cydonia, the location of the "face on Mars": A passion of its adherents, who want so badly to believe that intelligent civilizations once thrived on Mars that they cannot see, or are in denial of the evidence against it.

As for ancient civilizations, if you wish to go back 5,000 years to cite behavior you wish to emulate, then consider some other baggage that comes with it—the worship of cats, the divinity of the Pharaohs, the obsession with expensive, overbuilt, triangular tombstones. And while we are at it, why not include the Aztecs? Time to rip the pulsing heart out of virgins to appease the gods. Or time to eat the flesh of those you conquer, to make you stronger. And to complete the scene, why not die of disease and pestilence before you turn 40?

Like so many other attributes, astrology was not an achievement of these civilizations, it was a cultural liability.

Keeping it real,
Neil deGrasse Tyson

Psychic Teleportation

Saturday, November 6, 2004
Neil,
This is really stupid!!!

News Headline: "Air force report calls for $7.5 million to study psychic teleportation"

James McGaha, MS, FRAS*
Director, Grasslands Observatory

Hi James,
At $7.5 million, this represents ten minutes of military spending, out of their $400† billion annual budget. So perhaps we should ask, is there any fraction of the military budget that should be reserved for fringe investigations? I recently collected several pages of embarrassingly wrong quotes from people, many of whom should have known better, about what was or was not possible in the field of transportation. Here are a few examples:

* James McGaha (USAF Retired) is an acquaintance and fellow skeptic, with a long history of debunking claims that UFOs are visiting aliens. Here, I rein in his skepticism . . . just a bit.

† The US military budget for 2004 was $400 billion. Since then, the budget has increased to $600 billion, by far the largest in the world. Three times that of the next country (China) and more than the next ten countries combined.

"It is entirely impossible for man to rise into the air and float there. For this, you would need wings of tremendous dimensions and they would have to be moved at a speed of three feet per second. Only a fool would expect such a thing to be realized."

—Joseph de Lalande, Mathematician of the French Academy, 1782

"What can be more palpably absurd than the prospect held out of locomotives traveling twice as fast as stage-coaches?"

—*The Quarterly Review*, 1825

"Men might as well project a voyage to the moon as attempt to employ steam navigation against the stormy north Atlantic Ocean."

—Astronomer Dionysius Lardner, 1838

"Man will not fly for 50 years."

—Wilbur Wright to his brother Orville, 1901

"There is no hope for the fanciful idea of reaching the moon, because of insurmountable barriers to escaping the Earth's gravity."

—University of Chicago Astronomer, Dr. F. R. Moulton, 1932

Of course, these quotes, in the end, are about perceived limits to our technology and not the laws of physics themselves, but the public (who funds the military) does not make this distinction. What would it look like for a physicist to stand up in front of the Armed Services Committee of Congress and declare: "Do not spend a dime on psychic teleportation. It will never work," but simultaneously confess, "But yes, quantum teleportation is real."

In contrast with the Cold War era, today's Air Force is, by and large, frugal. They, for example, abandoned the space shuttle as a principal launch platform for their satellites, citing unreasonable costs compared with unmanned rockets. This, in spite of having influenced the original design of the shuttle itself in the service of their needs. Another manifestation of this frugality is that if they pay for a study, and find that some phenomenon or mechanism will not work, they will not pay for it again.

So I have no easy answer to this issue, except that dropping $7.5 million down to $0, on assertions of impossibility, comes at a social and political cost.

Neil

Parallel Universe

In the 1990s, while working backstage at a theater, Corinne experienced an unexplained phenomenon. She saw a male version of herself, looking straight at her, dressed the same, and walking in the same direction. He regarded her with no less fascination than she regarded him. Corinne assures me that she is psychiatrically stable, and in November 2008 simply wondered aloud whether she may have witnessed a portal to a parallel universe.

Dear Corinne,
Thanks for sharing this account.

I'm not much concerned about your psychiatric profile. Some famous scientists have been what many would call crazy. What matters is experiment, not eyewitness testimony.

Over the years, the methods and tools of science have shown us that, despite what some philosophers have claimed, a reality exists independent of our perception of it. We know this because, for example, the laws of gravity are there and work every time, no matter who is doing the experiment, no matter what apparatus you use to measure it, and they work whether or not you choose to believe in them.

Essentially every other claim to reality is overwhelmingly likely to be psychological rather than physical. (Ignoring, of course, hoaxes, other deceptions, or simple ignorance of natural phenomena.) This includes ghosts, apparitions, spirits, etc. None of these claims survive laboratory scrutiny. They all simply go away under controlled circumstances.

So if you really did see a parallel universe rather than have a mental apparition, then what you saw would exist independent of you, and should be measurable by everyone around you. But you do not have enough data to demonstrate this.

Next time this happens, be sure to conduct simple experiments on it:

- Can you communicate with it?
- Does it show up in a mirror?

- Did it leave fingerprints?
- Did other people see or interact with it?
- Was there a smell?
- Was there a sound?
- etc.

All of this would help to establish an existence outside of your head, if, in fact, your experience was real and not imagined.

In any case, next time, bring your camera. And perhaps a net.

Sincerely,
Neil deGrasse Tyson

Moons of Mars

In June 2005, Tom wrote from Canada asking about how the eighteenth-century English satirist Jonathan Swift could have possibly known that Mars has two moons when writing his classic Gulliver's Travels, *a full 160 years before their discovery. Swift gave details of their orbits around Mars. Might he have had access to some ancient form of knowing that we discount or ignore today?*

Hello Tom,
Thanks for your inquiry.

In the days of Jonathan Swift, Venus was known to have no moons, Earth one moon, and Jupiter four moons.

If Swift were to guess a moon sequence for these

planets, in order from the Sun, he would not choose zero or one or four. Those moon-counts were already taken. This leaves two or three for the yet-to-be-discovered moons for Mars, a planet that orbits between Earth and Jupiter. Given the choice, Swift chose two, as, I think, most people would.

Kepler's* laws of planetary motion were well known in the day. And they applied to Jupiter's moons in orbit around Jupiter as well as to the planets in orbit around the Sun. So Swift applied these laws to Mars' hypothetical two moons. But Swift had to assume orbital distances for them both. Upon doing so, a simple calculation gives a corresponding period of revolution for them. If you check his calculation you can verify that Swift did his homework—correctly.

But what many people failed to check was whether he got the right distances in the first place. He did not. In fact, he was way off, indicating that he, as suspected, had no premonitional clue about the actual moons of Mars.

If you are curious, the innermost moon, Phobos, orbits at 5,800 miles from Mars as opposed to 12,300 miles (3 diameters) as Swift said, and the outermost moon, Deimos, orbits at 14,600 miles from Mars as opposed to 20,500 miles (5 diameters).

Sincerely,
Neil deGrasse Tyson

* Johannes Kepler, a German astronomer and mathematician (1571–1630).

Perpetual Motion

In December 2008, Shawn wanted to run by me his ideas for a perpetual motion machine. He was confident that the laws of thermodynamics are not as sacrosanct as physicists claim, and that if oil companies find out about his idea, they will suppress the discovery. So Shawn sought my help to bring his invention forward, in ways that can change the world.

Dear Shawn,
The U.S. Patent and Trademark Office no longer accepts proposals for perpetual motion machines without a working model to demonstrate the invention. Why? Perpetual Motion machines violate long-established and well-tested laws of physics.

So if you have an idea for one, you cannot realistically expect any scientifically educated person to give it attention.

This leaves you with one, and only one option: to build it and demonstrate it. If the machine works as you say, then people will beat a path to your door.

Sincerely,
Neil deGrasse Tyson

Shawn replied with some attitude, declaring that at one time everyone was sure that Earth was flat, that the atom was indivisible, and that DC current was the only option to distribute electricity. And that maybe I was stubbornly close-minded in my reply.

Dear Shawn,

Many or your assumptions derive from an incomplete understanding of how science works. In the era of "modern" experimental science, which is basically traceable from Galileo and Sir Francis Bacon* onward (the past 400 years), there is tested science that has achieved consensus and there is frontier science. The frontier science changes monthly, if not weekly, awaiting good enough data to resolve the controversies. The tested science, that for which consensus of observations and experiment has emerged, *does not* change. What can, and often does, happen is that new ideas extend the range of previously tested ideas, but it does not undo them.

In your short list above, the flat Earth and the indivisible atom predate modern science. And crude oil in lamps and DC current are not tests of scientific principles. They are technological applications of science awaiting improvements upon them. But new technologies did not violate established laws of physics. They are (and continue to be) technological innovations happening within the known laws of physics.

Most importantly, the history of scientific discovery tells us that your quest is misguided, so the burden of proof is 100% on your shoulders.

Don't let me stop you. Like I said, go ahead and build the thing. If you succeed, you will have demonstrated a heretofore unknown law of physics. These are rare, but always welcome on the scene. And you will become rich and famous overnight.

* Sir Francis Bacon (1561–1626), English scientist, philosopher, and statesman.

And thanks for that "good luck," but in truth, I'm not the one who needs the luck here.

Sincerely,
Neil deGrasse Tyson

Dogon Predictions

Monday, July 30, 2007
Dr. Tyson, my name is Phil Dabney, a teacher from Lake Taylor High School, connected to Norfolk Public Schools. I met you today at the Physics Convention in Greensboro, North Carolina.

Thank you for the outstanding delivery you made today. I was particularly impressed with the "meet the students where they are at" approach when it comes to education. This may be the main reason why your books are so popular with all age groups.

Due to time constraints, I was unable to ask you a question concerning the Dogon's prediction of the star Sirius being a binary star before its confirmation via telescope. I believe this was well documented by two French anthropologists in the book, The Pale Fox.*

Would you comment on the veracity of this prediction?

Thanks for listening.

All the best,
Phil Dabney

* M. Griaule and G. Dieterlen, *The Pale Fox* (Baltimore, MD: Afrikan World Books, 1986).

Dear Mr. Dabney,
Happy to offer some perspectives on your inquiry.

As we know, the star Sirius, the brightest in the nighttime sky, was important to the Dogon tribe of Mali in West Africa, and to other cultures as well, including the ancient Egyptians, where the rising of Sirius in the sky just before the Sun (termed the "heliacal rising") signaled the time of year when the Nile would flood the valley bringing needed water to their desert climate. In fact, this event signified the new year of the Egyptian calendar.

Without the assistance of technology, it is physically impossible for the human eye to see the binary companion star to Sirius, known as "Sirius-B." Sirius-B's brightness falls below the light detection limits of the human retina. But more importantly, the huge difference in relative brightness between the two stars leaves Sirius-B lost in the glare of Sirius-A, much the way a firefly goes unnoticed in sunlight. Also, the separation between the two stars is smaller in angle than what can be resolved by the lens of the human eyeball. These limits are set by the physics of sight and not by the biology of the individual.

Sirius-B was discovered in 1862. At the time, two things were true: the event was widely publicized, with page-1 news stories all over Europe. And at that time, European missionaries, explorers, and imperialists were common all throughout Africa. And lastly, the French anthropologists you refer to came upon the Dogon *after* the discovery of Sirius B.

These are the fundamentals of the case. The Rutgers University historian and anthropologist Ivan Van

Sertima has written about the Dogon* and invokes heavy doses of speculation in his effort to credit the Dogon with Sirius-B's discovery. This includes the specious claim that the sunlight-absorbing power of melanin in the skin of Black Africans imbued the Dogon with heightened powers of perception.

So either the Dogon had pre-cognition of Sirius-B via some mysterious, and still undiscovered way of knowing, unique to them in the world, or another European visitor (anthropologist or otherwise) came across the Dogon *before* the French ones did, saw that the star Sirius was in their culture, and shared with them the highly publicized recent discovery of Sirius-B, yet did not write about this encounter. The Dogon immediately adopt into their culture this extra information about their favorite cosmic object, and the French anthropologists come upon them later, startled by the Dogon's detailed knowledge.

Furthermore, if you read other elements of Dogon culture and stories of nature, none of them have the precision of information that Sirius-B enjoys within their culture. Their stories are romantic and poetic and ring just the way creation myths do from most cultures.

While we do not know for sure whether the Dogon were visited by informed Europeans before the French anthropologists got there, the evidence

* Ivan Van Sertima, *Blacks in Science: Ancient and Modern* (Abingdon-on-Thames, UK: Transaction Books [now Routledge], 1991).

strongly suggests this. Any other supposition is carrying Afrocentrism further than the data warrant.

Thanks for your inquiry.

Best to you,
Neil

Bigfoot

In January 2008, Alex sought my view on whether a large hairy ape could actually be roaming the Pacific Northwest.

Dear Alex,

At a time before the world was fully mapped, European explorers told great tales of new plants and animals, especially in their journeys through Africa and Asia. They collected what they could and brought it back for research and display in museums. New large animals were being identified frequently. This was the beginning of Natural History as an academic discipline.*

After all landmasses of the world were mapped and settled, the discovery rate of new, exotic creatures dropped precipitously. This offers a strong indication that all large (land) animals are known and documented. The new land species that are discovered

* Of course natural history as a topic of interest goes way back. The Roman author and naval commander Pliny the Elder wrote a book titled *Natural History* (ca. AD 79) that compiled all ancient knowledge of the natural world into one volume. And Leonardo da Vinci (1452–1519) himself was also an acute observer of nature.

annually today tend to be small animals or mild variations (e.g. sub-species) among well-documented ones.

Occasionally large sea creatures are found, but this is understandable since we do not live in the sea, and do not persistently monitor the sea floor for life.

So the likelihood that a large (land) animal has gone unnoticed in modern times is near zero.

Neil deGrasse Tyson

Alex replied with great skepticism of my closed-minded views, while reminding me that the Pacific Northwest contains three million acres of unmonitored forest. He further noted that one cannot discount every sighting of a large hairy ape as nonsense. And as such, there must be something to it all.

Dear Alex,

I commented on the rate of discovery and documentation of large animals. Sightings without tangible evidence (such as a dead body for the lab, or hair, for example) do not constitute discovery. Eyewitness testimony is well known among psychologists and scientists to be the *least reliable* form of evidence of them all. Hence, it is strongly discounted, as the researcher patiently awaits the presentation of tangible evidence in support of the extraordinary claim.

Note that every sighting may be real. But without a body, or other firm evidence that does not pivot on human perception, the claims are useless to the research scientist.

FYI: "useless" is not the same word as "false."

Until somebody produces biological tissue (even big-foot poop would be a start) from which DNA can be extracted, there is not much a biologist can do with the claims.

If you feel confident that there are eight-foot, undocumented, prehistoric apes running around the Pacific Northwest, you should mount expeditions to find them—you don't have to kill it; just capture one.

Your efforts will be better spent in search of useful evidence than in trying to convince people of what you think is true in the absence of it.

Neil

Sixth Sense

Tuesday, February 6, 2007

Dear Doctor Tyson,

I'm reading your book Death By Black Hole.* *First let me say that your writing style is just like your speaking style—clear, understandable and pleasant. I just miss the laughing I hear when you are interviewed live. Second, I must comment on what you said about the sixth sense.*

The headline we never see is:
Psychic Wins Lottery Again.

I witnessed my grandmother, who had "the gift" and used it just like her other senses. She knew when people were coming to visit, so she changed all the beds and bought more groceries. She knew whether or not my father was going to

* Neil deGrasse Tyson, *Death by Black Hole: And Other Cosmic Quandaries* (New York: W. W. Norton, 2007).

be home for dinner or not and set the table accordingly. She would wake up when the cow was calving and bake pies when visitors were coming. All of this was a sense to her—not a "Psychic Hotline" kind of thing, but an extra perception that she took in stride just like her other five senses. She was from Ireland and her grandmother was the same way. It was how they lived.

My father could always tell when women were pregnant, often even before they knew themselves (and no, it wasn't because he made them that way). Perhaps that had to do with micro-changes or pheromones, but he did it all the time.

Anyways, I'm sure you have heard lots of these kinds of stories. From my own observations, the only kind of sixth sense I believe in is a rather primitive kind of feeling that helps us get along.

Thanks for all the work.
Kathleen Fairweather

Hello Kathleen,
Thank you for your testimonial. I will not stand in denial of your account of your parents' powers of perception.

But in every case where these powers are studied in the lab, the powers fail—or more precisely stated, the people who claim to have such powers do no better than chance in experiments designed to test them. Decades of articles in the magazine *The Skeptical Inquirer** have documented this.

* *The Skeptical Inquirer,* a bimonthly magazine published by the Committee for Skeptical Inquiry (Amherst, NY).

So either the power goes away under controlled circumstances or the people are remembering the hits and not the misses—one of the most common perception failures of the human mind. A well-studied phenomenon by psychologists, it covers, for example, people who have premonitions about a friend's health. You call, only to discover that your friend is in the hospital, or not feeling well.

When these events do happen, they become powerful memories that supplant one's recollections of failures. Like I said, there is an entire literature on this, which I cannot review here. But the experimental method, telling us more about ourselves than we can determine on our own, is what enabled society to move from the days of superstition and witch burnings (where women were believed to have ungodly powers) into the era of empirical inquiry, birthing the industrial revolution and modern life.

All the best in your pursuits of mind,
body, and spirit,
Neil deGrasse Tyson

Chapter 3

Musings

Random things people think are actually a category unto themselves.

Complexity

Friday, March 8, 2019
Hello Guru:
I recently saw a daddy-long-legs, which reminded me that he and I have a common ancestor way way back.

Our huge subsequent divergence was triggered by many, many trillions of random DNA mutations and helix expansions, of which some 3 billion nucleotides survive in every one of my trillions of cells. And those 3 billion must be in the single correct sequence to make me, develop me, run all my physiologies, and even dictate my instinct.

How could a mere 3 gigabytes do all that. It takes far more just to run my iPhone.

Those 3 gigs don't seem to be enough to just dictate how my brain's 100 billion neurons and their trillions of synapses behave.

My religious friends have an easy answer which I don't accept.

Best wishes,
Josh S. Weston

Josh,
Simple sets of "rules" can lead to extraordinary complexity.

For example, people in a capitalist society generally value money. Add to this a few simple economic tenets, such as "buy something and sell it for more," coupled with a basic understanding of "supply & demand" and lo and behold, you have a corner grocery store that offers ten varieties of milk, obtained from farms hundreds of miles away, delivered to you in a supply chain via refrigerated trucks, and available to you 24/7.

You can say that the creator of the universe has your health as a priority, and established this highly complex system at every step, just to make sure you drink fresh milk daily. Or you can say that greed drives it all.

But wait, there's more . . .

How about the fact that the entire universe is composed of just 92 elements?

Or that there are only four fundamental forces of nature? (strong, weak, electromagnetic, gravity)

Or that there are only four classes of fundamental particles? (quarks, electrons, neutrinos, and photons)

Or that nearly all behavior of electromagnetic

waves (light) can be derived from a set of four equations that all fit on a Post-it® note?

So you can be awed by the complexities manifest the world, or you can instead be astonished at how simple it all is.

—Neil

Spirals

Paulette B. Cooper described herself as math challenged. In spite of this, she could not help but notice the ubiquity of spiral shapes in the universe, from galaxies to hurricanes to the spiral-generating Fibonacci sequence. * *She wrote in March 2006 to see if they were all connected in some cosmic way.*

Hello Paulette,

One of the great challenges of cosmic discovery, and discovery in general, is knowing the difference between when things *look* the same and when they *are* the same.

Spiral galaxies and hurricanes have nothing whatever to do with one another, in spite of their similar appearances. Additionally, you can have a two-armed spiral galaxy, but a two-armed hurricane has never been observed.

* Fibonacci (ca. 1170 – ca. 1250), Italian (Pisa) mathematician, most famous for the number series that bears his name, where each entry is the sum of the two previous entries; i.e., 1, 1, 2, 3, 5, 8, 13, 21, 34, etc.

More importantly, the forces that are at work in the two regimes are entirely independent of each other. Those that make hurricanes involve pressure differences in the atmosphere, the heating of oceanic waters, and the Coriolis force which pushes clouds sideways, making the circular patterns you see. In a galaxy, the relevant forces are entirely gravitational, and the spiral pattern is traced by newborn stars.

Consider other things that look alike. When Sir William Herschel, in the 1800s, first saw dots of light that moved slowly across the sky, he knew they could not be stars, but they looked like stars through his telescope, so he called them "star-like," which in Latin becomes "aster-oid" or just asteroids. Their similar appearance through a telescope eyepiece was irrelevant to what they actually are. Stars are billions of times larger than asteroids and operate under different forces of nature.

Another attempt to say things were the same (but were not) occurred with the early ideas of what atoms might be—imagined as mini solar systems, with the nucleus as the Sun and the electrons in "orbit" around it. Early textbooks show pictures of this concept. But the laws that describe the atom have nothing whatever to do with the laws that describe planetary orbits. Not only that, the analogy left a misleading imprint on the vocabulary of atomic physics. For example, we describe electrons as occupying "orbitals," even though their paths are best described by "clouds."

So yes, appearances can be deceiving, and it's

always best instead to ask, "What is the thing?" rather than "What does the thing look like?"

Sincerely,
Neil deGrasse Tyson

Roots

In February 2014, Henry Louis "Skip" Gates Jr., a professor of African and African American Studies at Harvard University, invited me to participate in his PBS series Finding Your Roots, *in which he explores the genetic heritage of notable Americans. The goal of the series is "to marry top-notch genealogy with cutting edge genetic science to reimagine the meaning of race." The impressive list of celebrities who had already participated included Martha Stewart, Oprah Winfrey, Mike Nichols, Samuel L. Jackson, Barbara Walters, and Chris Rock. I declined the invitation.*

I happen to know Professor Gates personally, via not-for-profit boards on which we have served, and so I was candid in my reply.

Hi Skip,
Thanks for this invitation to participate in your immensely successful series. It's so widely embraced and talked about.

For me, however, I have an unorthodox philosophy regarding root-finding. I just don't care. And that's not a passive absence of caring it's an active absence of caring. Since *any two people in the world* have a common ancestor—depending on how far back you look, the line we draw to establish family lineage is entirely arbitrary.

When I wonder what I am capable of as a human being, I don't look to "relatives," I look to all human beings. That is the genetic relationship that matters to me. The genius of Isaac Newton, the courage of Joan of Arc and Gandhi, the athletic feats of Michael Jordan, the oratorical skills of Sir Winston Churchill, the compassion of Mother Teresa. I look to the entire human race for inspiration for what I can be—because I am human. I don't care if I am a descendent of kings or paupers, saints or sinners, the brave or cowardly. My life is what I make of it.

So I respectfully decline your invitation, but I do so knowing that ever since Alex Haley, most people find this pastime to be immensely enlightening. And I will not deny them this insight and revelation to their past. So I mostly keep these sentiments to myself.

Continued success with the series.

Best, always,
Neil

BC / AD

In April 2009, Lionel, an ardent atheist, expressed his frustration and disapproval at being forced to reckon calendar*

* *a-theist*: literally, "without God." Never liked the term. Odd for words to exist that tell you what you are not. Is there a word for non-golfers? Non-chefs? Non-astronauts?

time on religious foundations, specifically Christian traditions. He wanted science to come up with a more sensible system of time given what we know today about the age and origin of Earth and of the universe.

Dear Lionel,
Thanks for sharing your thoughts on this matter, and for inquiring about mine. Several points to consider:

1. Most time reckoning for the distant past of Earth and the universe does not reference a particular calendar at all. It simply counts years before the present. For example, nobody says Earth was formed in 4.6 billion years BC. We simply say it formed 4.6 billion years ago. Same holds true for geological and biological time-keeping.

2. The origin of Earth spanned at least a hundred million years. So a precise date and time to start a cosmic calendar would be meaningless—like attempting to celebrate the nanosecond (billionth of a second) that you were born. The time it took you to exit the birth canal vastly exceeded this measure of things. So we sensibly record births only to the nearest minute.

3. For times and dates within recorded history, the Christian Gregorian calendar is the international standard, in which BC, written following the year, stands for Before Christ, and AD,

written preceding the year, stands for Anno Domini, Latin for the "Year of our Lord." Other calendars exist—Jewish, Muslim, Chinese, etc., in which they each link the zero-point of their calendar to some event in the respective religion or culture. But those calendars today are more ceremonial than practical.

4. The Gregorian calendar is, quite simply, the most accurate and stable calendar ever devised. The Jesuit priests, appointed by Pope Gregory back in the 16th century, did a bang-up job in their calculations. They corrected the failing Julian calendar, in which the spring equinox had back-drifted over the centuries from the familiar date of March 21, to March 10. The correspondence of the spring equinox with March 21 is forever secured, never drifting more than a single leap day in one direction or another. Meanwhile, other systems, especially lunar calendars, intermittently need to introduce entire months to reconcile the timekeeping with Earth's location in orbit around the Sun.

When you do something right, and you do it better than anybody else before you, you get to name it. Any atheists of the day were not into calendars. Of course, they've never really been into calendars. So atheists are out of this loop, except for the introduction of BCE and CE—Before Common Era (replacing BC) and Common Era (replacing AD).

Consider that Handel's *Messiah* is among the greatest choral works ever composed. So too is Bach's *Mass in B-Minor.* Yet neither of these works would exist were it not for somebody inspired by Jesus. This does not (or at least should not) subtract from the brilliance, the beauty, and the majesty of these great works of music.

Furthermore, as an atheist, you surely use the words: "holiday" and "goodbye," yet their origins come from "holy day," and "god be with you."

As in anything in life, you must choose your battles.

So may I recommend that you adopt CE and BCE and leave it at that? Instead, devote your energies to the real battlefield—the "sanctity" of the science classroom in the face of religious fundamentalists who perennially attempt to influence the science curriculum with religious-based philosophies.

Sincerely,
Neil deGrasse Tyson

Skies Over Iraq

Monday, March 5, 2007
Dear Neil,
My name is Derrick Philips. I am a Private First Class in the Army. I am currently deployed outside of Balad, Iraq. I asked my wife to send me a copy of your newest book and since I received it two days ago, I have not been able to put it down. Currently, I am charged with the mundane task of pulling guard duty. After my grueling 12-hour day of

watching mule drawn carts and standing in one spot, I get an opportunity to sit down and open your book, Death By Black Hole. *I am sure that a lot of people are enjoying it in as many ways as there are people reading it. I think that you might like to hear my unique way of deriving pleasure from it.*

*I am about an hour north of Baghdad. This is a place in your book that has gotten a few references.** *I am able to talk to some of the local people who are well aware of their past significance in the role of the sciences and give them opportunity to speak to me of many other things that I was previously unaware of, that have transpired right here in my temporary back yard. These conversations, that are spawned from the knowledge imparted by your book, leave me feeling more like a heavily armed tourist, rather than an occupying invader.*

A book like yours, I imagine, leaves people wanting to gaze upon the stars and reflect on some of what they have just read. I have discovered that the night sky is teeming with a lot more than I ever thought possible, when I turn on my Night Vision Goggles. How many of your readers can say that they were inspired to use defense technology to relax after a long day at work? Well, some I imagine.

Anyhow, your book has inspired me to think, to use my head for something other than just an ear spacer! I have to thank you for helping to stave off my boredom for part of the year that I am here.

I am highly interested in this subject, but seriously under-educated. I am doing independent research into our cosmos so

* During the Golden Age of Islam of a millennium ago, the intellectual center of the world was Baghdad.

that I can relay this information to my children. They seem to share my fascinations, and with a minimal investment of a telescope I can see a lot of serious quality time coming from that endeavor.

In short, I just wanted to write to you and thank you for the contribution that you have made to MY war effort.

Sincerely,
PFC Derrick Philips

Dear PFC Philips,
Thank you for your kind words about my latest book and its relationship to your service in Iraq. It's an honor to help your time pass.

With regard to your night scope—connections between astrophysics and the military extend far beyond most people's awareness. I am working on a book now that highlights the countless connections that exist between them.*

And yes, Baghdad has deep history in the sciences, especially in mathematics—algebra in particular. Furthermore, next time you look up at night (as you may recall from the book), two-thirds of all stars that have names, are named in Arabic, enabled by major advances in navigation a thousand years ago.

The most enduring thing about being human is the discovery of cosmic truths that transcend cul-

* Neil deGrasse Tyson and Avis Lang, *Accessory to War: The Unspoken Alliance Between Astrophysics and the Military* (New York: W. W. Norton, 2018).

ture, politics, religion, and time, forming the corpus of knowledge and wisdom that we call civilization.

All the best, on Earth and in the universe.

Neil deGrasse Tyson

Seeing Stars

The "Metropolitan Diary" is a weekly feature of the New York Times *in which readers share unique tales of city life, as I did back in 1993.*

Wednesday, December 15, 1993
New York Times

Dear Metropolitan Diary:
An elderly woman with a strong Brooklyn accent recently called my office at Columbia University's Department of Astronomy to ask about a bright glowing object she saw "hovering" outside her window the night before. I knew that the planet Venus happened to be bright and well-placed in the west for viewing in the early evening sky, but I asked more questions to verify my suspicions. After sifting through answers like, "It's a little bit higher than the roof of Marty's Deli," I concluded that the brightness, compass direction, elevation above the horizon, and time of observation were indeed consistent with her having seen the planet Venus. Realizing that she has probably lived in Brooklyn most of her life, I asked her why she called now and not at any of the hun-

dreds of other times that Venus was bright over the western horizon. She replied, "I've never noticed it before!"

You must understand that to an astronomer this is an astonishing statement. I was compelled to explore her response further. I asked how long she has lived in her apartment. "Thirty years." I asked her whether she has ever looked out her window before. "I used to always keep my curtains closed, but now I keep them open." Naturally, I then asked her why she now keeps her curtains open. "There used to be a large apartment building outside my window but they tore it down. Now I can see the sky and it's beautiful."

Neil deGrasse Tyson
Manhattan

Lucy in the Sky with Diamonds

Wednesday, June 10, 2009
I'm Georgette Burrell and I'm seven years old. I saw your special on how Pluto is a dwarf planet. I thought that it was very cool. I heard of a planet (or star) called Lucy that is a big diamond. My question is how would scientists know about what it is if it's so far away?

Thanks,
Georgette

Excellent question, Georgette.

Many dead stars are made of carbon (they are white dwarfs). Under high pressure, pure carbon turns into diamond. These stars have strong gravity, which puts their carbon under high pressure. So we can use math to calculate the star might just be made of pure diamond.

Neil

I'd Rather Direct

Tuesday, July 22, 2008

Dear Mr. Tyson,

I am a member of the Writer's Guild of America, currently working on a script about interplanetary travel. I have seen you on numerous TV specials and I am a big admirer of your frankness in observations. One that struck me, and in turn made me seek your counsel, was your matter-of-fact summation of what if something goes wrong in space—your answer: "You die!"

I am working on a script about an astronaut who is sent off to Japetus, the third largest moon of Saturn, to investigate a mysterious alien transmission from the surface. Here's the thing—I want to get it right. I want this to be as accurate as possible. Could you answer a few questions about the inherent dangers of such a long space travel, from both inside and outside the ship?

Sincerely,
Andrei Anson

1. *How long would it take to get to Japetus?*

Dear Andrei,
Thanks for your questions. How long to Iapetus? Any amount of time you choose. An energy-minimizing ballistic trajectory would take about ten years. But if fuel is no object, then you can accelerate for most of the trip, then use fuel to decelerate—this gives you artificial gravity en route, and you could get there in a year or two.

2. *In my script, I have a new shuttle meeting up with the space station to affix its new fuel supply. How would they actually get to Japetus? Slingshot, etc.?*

In Earth orbit, you already possess half the energy to get anywhere else in the solar system. In other words, the energy necessary to achieve Earth orbit is exactly half the energy it takes to leave Earth entirely. Slingshot maneuvers are for spacecraft that do not launch with enough fuel to reach their destination. They take longer than ballistic trajectories because their total distance traveled can be twice what it would otherwise be, as they fall towards the planets and moons that would provide the needed gravity boost.

3. *How fast could we have them going? Is 39,000 mph the realistic limit these days?*

Any speed at all. It's just a matter of fuel available to accelerate and decelerate. Escape velocity for Earth is 25,000 mph. That should get you to Iapetus in 10 years.

4. *How would they get back?*

You need even more fuel to get back than to get out there. You'd have to refuel at Saturn somehow. Saturn's atmosphere contains molecules that you could use for this purpose, including water. But you would need a factory that could separate the H from the O in H_2O. Then the two elements can be combined in a rocket motor, to form rocket fuel. Or they could just get fuel from an alien fillin' station.

5. *What would happen to our hero, Tom, if he were to be stuck in space for the next twenty or so years? Physically speaking.*

Nothing. Unless he runs out of food.

6. *And lastly, how could I cripple the ship permanently?*

One scenario: you use Saturn's atmosphere to aerobrake (see the 1984 film *2010: The Year We Make Contact*), but a hole develops in the fuselage and hot air leaks into critical engine parts that permanently damages the throttle controllers and fuel tank. You have fuel initially. You have rocket motors. You just have no control of how much they fire. And then all the fuel leaks away as he spins off into oblivion.

7. *I was thinking some debris from an asteroid that just misses the ship. But its nearby debris does cause damage, rendering the ship useless. Or, quite frankly, is there some kind of other way to strand them? I know at 39,000 mph, most things would just probably obliterate the ship, something I do not want to do.*

Unlikely. Asteroids are few and far between. Or rather, space is vast, so they are actually many and far between. You could instead slingshot the ship around Saturn, en route to Jupiter, but then it accidentally gets catapulted through the field of Trojan Asteroids, which Jupiter has gravitationally trapped in orbit around the Sun. The ship then suffers collision damage that they are unable to repair—perhaps with lost fuel. Then requiring the aerobraking maneuver around Saturn, since the ship would not have enough fuel to decelerate. This ensures that Major Tom does not leave the solar system altogether, and dies in orbit around Saturn.

Have a nice day!
Neil deGrasse Tyson

Worst Ever

Wednesday, July 8, 2009
Dear Mr. Tyson,
*I just wanted to know: what do you think is the absolute worst cinematic offender to science? To make it easy, we will exclude all movies of the pre-*2001: A Space Odyssey *period, so you don't have to name Ed Wood's filmography. And what about the movie* Armageddon? *That movie sucked on both a scientific and artistic scale.*

Anyway, I hope you have time to reply, since I know you are busy, but I am unfortunately a very curious creature. I thank you for your time. Keep on rockin' in the free world.

Chris Bostwick

Dear Chris:
Disney's 1979 film *The Black Hole*. Worst ever—given how fertile the scientific material was—until the 1998 film *Armageddon*, which violates more laws of physics (per minute) than any other film in the universe.

Neil

A Viral Blunder

Tuesday, January 8, 2019
Dear Dr. Tyson,
We'd like to begin by introducing ourselves—Samyuktha and I are two NYC medical students and, more importantly, museum aficionados who frequent the American Museum of Natural History. We are writing to inform you about a small but substantial inaccuracy on one of the displays. Regarding a rhinovirus, the board reads "Rhinoviruses are among the major causes of the common cold. They consist of DNA surrounded by a protein coat." However, a rhinovirus contains RNA (ribonucleic acid) and not DNA (deoxyribonucleic acid) as the board states.

We acknowledge that this may seem trivial and do not mean to sound nitpicky, but whether or not a virus contains DNA or RNA is actually one of the most fundamental ways in which viruses are classified and distinguished. It affects the viruses' mode of transmission, mode of replication, stability, and physical characteristics among other core properties. Therefore, we decided this was important enough to write to you about.

Sincerely,
Samyuktha Guttha & Aneek Patel

Dear Samyuktha & Aneek,
Everybody knows that viruses carry RNA and not DNA—except, apparently, all of us who wrote and reviewed the panel text as well as the tens of millions

of people who have seen this exhibit since opening day, 247 moons ago.

I even checked my original files to see if there was a transcription error. That way we could blame the fabricators of the exhibit instead of ourselves. But, alas, it was the submitted text that had the error.

So . . . where were you both, twenty years ago when we mounted that panel? Could have used you back then!

Thanks for the sharp eyes.
And we'll get right to fixing that text.
Neil

Samyuktha Guttha and Aneek Patel replied . . .

Thank you so much. We were around 2 years old back then, but you definitely should have reached out!

Breaking Up Is Easy to Do

An open letter to all my fellow employees at the American Museum of Natural History.

Thursday Afternoon, May 4, 2006
Dear Museum Community,
As you may already know, "countless" comets (perhaps as many as trillions) orbit the Sun, along with everything else. The public will typically hear only about those bright enough to see with the unaided eye, or, of course, those that are about to hit something.

Unlike the near-circular orbits of planets, most comets travel on highly elongated trajectories, crossing other people's orbits as they plunge in and out of the inner solar system. Made principally of ice, as a comet nears the Sun's heat its outer layers evaporate, creating an enormous ball of reflective gas—the "coma." These gases also stretch through interplanetary space, forming the celebrated comet's "tail."

We have a good idea what comets are made of, but we do not know how solid they are. The range of structural integrity among the solar system's comets is surely wide, just as some snowballs will stick well, while others will crumble the moment they leave your hand.

Just now becoming visible to the naked eye, comet Schwassmann-Wachmann 3 in ten days will come within seven million miles of Earth—thirty times the Earth-Moon distance. From the stresses of being a comet this time around, its nucleus has begun to disintegrate, laying bare dozens of smaller chunks of ice, *each creating its own mini-coma and mini-tail.* On the sky, the lead comet fragment and its tail now span several degrees in angle, which is five or six times the width of the full Moon. Have a look at this stunning image of the comet, taken with the Hubble Space Telescope two weeks ago.

Visibility from light-polluted cities will be poor. But if you live in rural areas, Schwassmann-Wachmann 3 should be an easy target for you, with or without binoculars. This week, it moves across the constellation Hercules. Next week, it crosses the adjacent constellation Lyra. If you face south, both constellations will sit high in the sky, several hours before sunrise.

And contrary to the announcements of doomsday websites and heavily forwarded email alarms, this comet poses no harm to Earthlings.

As always, keep looking up.
Neil deGrasse Tyson

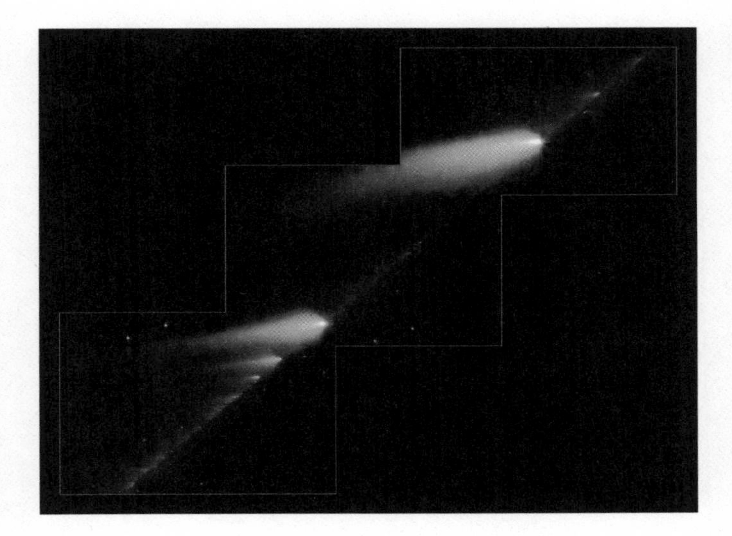

II.

Cosmos

The universe seen
as a well-ordered whole.

Chapter 4

Hate Mail

About a third of all correspondence I receive qualifies as fan mail. Occasionally, the opposite arrives in my inbox.

An Apology

Monday, June 18, 2012
Dear Dr. Neil deGrasse Tyson,
I am writing to you to deeply apologize for a nasty but stern coloring I sent to you, 12 years ago, when I was 10-years-old, calling you a "big poo-poo head" for demoting Pluto as a planet. * *Please accept my sincerest apologies, for I am a huge fan of your work and deeply regret hurting your feelings with such vicious and harsh words!*

Sincerely,
Michael C. Hotto

* While I did not personally demote Pluto from planet status, I was definitely an accessory to the indignity. This made me a public enemy of elementary school children nationwide.

Dear Michael,
I have only a vague memory of that particular letter—my file cabinet brims with such correspondence. But I nonetheless warmly accept your apology, knowing that you were simply being honest about your feelings at the time.

Sincerely,
Neil

An Appeal

Fall 2006. A third-grader writing from Peters Elementary School, Plantation, Florida.

Dear Scientest,
What do you call pluto if its not a planet anymore? If you make it a planet agian all the Science books will be right. Do poeple live On Pluto? If there are poeple who live there they won't exist. Why can't Pl.u.to be a planet? If its small doesn't mean that it doent have to be a planet anymore. Some poeple like pluto. If it doesn't exist then they don't have a favorite planet. Please write back, but not in cursive because I can't read in cursive.

Your friend,
Madeline Trost

At the time this letter arrived in my office at the Hayden Planetarium I was busy fielding hundreds of such letters and did not reply. But had I done so, this is what I would have said:

Dear Madeline,
If anybody is living on Pluto, I assure you they still exist, even after Pluto's demotion to Dwarf Planet status. So no need to fear for their lives. Also, if Pluto is anybody's favorite planet, then it can simply become their favorite Dwarf Planet. No harm there. But in any case, you are right about the textbooks. They will all have to be changed. Bad for book buyers. But good for publishers—they get to sell you the book again.

And here is my actual signature in cursive. It says Neil D. Tyson. Gotta start somewhere.

Your friend,

Neil D. Tyson

Moon Lover

Friday, January 6, 2007
Dr. Tyson,
Heard you on the Radio this morning, repeating the standard shibboleths about both Mars and the Moon. It is particularly disappointing, in your case, listening to you bashing the Moon when, as an astrophysicist, you know better than most that telescopes deployed on the dark side of the Moon would far outstrip any other means of studying the cosmos, most

assuredly including the Hubble telescope—placing equipment on the lunar surface, we could not dream of and would have no need to put anything into orbit at far higher costs.

The Moon is humanity's stepping-stone to the next leap in our evolution. Making us into a true space faring species, which will result in a very new and better understanding of who and what we are, and our shared destiny. When I look up in the sky, Dr. Tyson, and see the full Moon so bright and so close I can almost touch it, I certainly don't think of some dead useless mass floating in the heavens. When I look upon the Moon, I think of it being the year 2050 or 2075 with lights twinkling all over the lunar surface—the clear evidence that a new society is rising there, transforming humanity down here on Earth.

Best wishes,
Arthur Piccolo

Hello Mr. Piccolo,
Thank you for your candid remarks. Allow me to restate some issues for which there is broad consensus in the scientific community.

1. Without atmosphere or any history of running water on the Moon, or any likelihood of substantial quantities of water within the Moon (like aquifers, etc.), or any likelihood of life as we know it—or can imagine it—given its collision-induced formation mechanism, there is no debate about whether the Moon is dead, compared with Mars.

2. The Moon's primary scientific interest is geological and not chemical, biological, or astrophysical, unlike for Mars, where it is all of the above.

3. The scientific returns to astrophysics from our presence on the Moon will be meager, given the cost to get there, as discussed at length in a recent workshop I attended that explored this very subject, titled "Astrophysics Enabled by the Return to the Moon." You can Google the title. Radio telescopes on the far side (FYI: there is no permanent "dark side") led the list. And a few other interesting projects got people's attention. But overall, we will be piggybacking exploration missions because we can, not because we have made it a priority. And the greatest gains to astrophysics may simply be access to space-based architecture with no direct relevance to lunar surface activities.

4. Proof of liquid water in Mars is not what matters here. It's that the evidence points to this conclusion, and that is enough to justify further investigation. For if it's true, then the chances of life as we know it on Mars grow exponentially.

I respect your love for the Moon, but the depth of that love does not change the Moon's ranking as an object of scientific interest among the multidisciplinary community of research scientists.

Again, thank you for your interest.
Neil deGrasse Tyson

We Suck at Science

Thursday, July 5, 2012
To AMNH General Mailbox*

I was very saddened to read Neil deGrasse Tyson's tweet yesterday, Independence Day:

On the day we reserve to tell ourselves America is great - July 4 - Europe reminds us that we suck at science. #HiggsBoson

10:39 AM · Jul 4, 2012 · TweetDeck

9.6K Retweets **1.8K** Likes

Tyson has been serving American and world science well, even coming into his own as a spokesperson. What seems an attempted joke is disappointing as a spokesman derides and mocks a nation of scientists, especially while serving at the public platform of the Museum. While his insulting statement was made from a private account, he should not

* The American Museum of Natural History, New York City.

represent the American Museum of Natural History as listed in his bio.

Thank you for your time in answering my concern.

Jeff Provine

Dear Mr. Provine,

Thanks for your note of concern. I have several reactions to the issues you raise. I tend to be candid in personal correspondence, so I hope you view it as refreshing rather than abrasive.

1. Every metric of America's performance on the world stage of science, technology, engineering, and math (STEM) fields among industrialized nations put us in the bottom 10%. We also have an ever-growing fraction (nearing 50%) of the electorate that denies the discoveries of science when they conflict with their politics and/or religion. So to imply that I have somehow misrepresented the state of American science is simply false.

2. The back story here, known well to people who follow my writings, is that we began construction of the Superconducting Supercollider in the 1980s. That machine was designed with 3x the power of the Large Hadron Collider in Switzerland today, which is garnering all the physics headlines. Congress

cut the project entirely in the early 1990s, crippling particle physics in America. That's why we are bystanders and not leaders in these international headlines. And all this feeds the potency of the Tweet itself.

3. Your note implies that the Tweet may have somehow done a disservice to science or science education or to AMNH itself. That assumes, in part, that others feel the same as you do about that same tweet. But I have data on that very question. The Twitterverse keeps full track of replies, responses, re-tweets, etc., for every Tweet that gets posted. Within twelve hours, that tweet was re-posted (re-tweeted) nearly 12,000 times. This number is far and away (factor of three) the largest reposting of any of my previous 2700 tweets in the last three years. So the resonance was (and continues to be) very high, and does not happen to align with your concerns.

4. None of this is to say that your sentiments are not conceived out of deep care for our country. All it says is that your feelings are not representative. And so I'm left with the question of whether I alter what I am doing to satisfy the few, or keep what I am doing, which satisfies the many, drawing more and more people to hear about science.

5. Of course that which is right in the world is not always (hardly ever) a popularity contest. Principles can, should, and do matter without regard to sheer numbers of adherents. But I assert strongly there is no principle violated here at all. I am compelled to change or atone or apologize or adjust only if what I say is wrong or misleading or libelous, but not if what I say (Tweet) captures a deep truth that requires national action to amend.

Sincerely,
Neil deGrasse Tyson

I'm Not Paying!

Friday, May 16, 2008
Via Email to RNASA Organizers*

I hated every minute of that award acceptance speech by Dr. Neil deGrasse Tyson. I like him, and I like watching him on shows on the Science Channel, but I don't like his approach to funding the space program.

If space exploration is so great and so profitable, then why can't it exist without stealing money from me at the point

* Rotary National Award for Space Achievement, conferred at an annual black-tie banquet in Houston, Texas, the city that is the heart of America's manned space program. This letter-writer did not attend but viewed my acceptance speech on-line.

of a gun (taxes)? Why can't it just sell all of its innovations and exist without being a socialist program?

He says the Cassini mission to Saturn costs as much as Americans spend on lip balm . . . well . . . I freaking chose to buy the lip balm; you forced me to pay for your lame spacecraft. Whoever is prepared to give up their lip balm are the people that should VOLUNTARILY pay for space travel. Don't make me pay for it! Then maybe the people/companies that voluntarily pay for it can also be the ones that receive free, innovative, high-tech help from NASA.

Space exploration, funded by socialist means, at the point of a gun, makes this country worth defending? That kind of crap is what makes this country worth abandoning—it's the opposite of freedom.

Comparing us to China and the rest of Europe, and their innovations—is that what we want to become? Socialists and Communists, like they are?

Our country didn't become great because of socialism and big government; our country became great because of our relative freedom, in spite of big government and socialism. The thing that's ruining our country now is not freedom. What's ruining us now is big government and socialism. And socialists like Dr. Neil deGrasse Tyson, trying to get the government to steal more money from me to fund his pet program.

I mean, what if enough losers decided that it would be a good idea to steal tax money from YOU in order to fund a program to teach everyone in the U.S. to speak Spanish, because they have a bunch of reasons why it's good for all of us to learn to speak Spanish and English? You wouldn't like it, would you? THAT'S HOW I FEEL ABOUT SPACE EXPLORATION!

I like space exploration, and it's probably a pretty good thing—just don't force me to pay for it.

Adam Dirkmaat

Dear Mr. Dirkmaat,

Thanks, first, for spending the time to watch my acceptance remarks after being given the RNASA Space Communicator Award this past April in Houston. And thank you for sharing your impassioned view on government spending on space exploration.

You comment that the American space program is the product of a kind of tax-based socialism forced on disinterested Americans such as yourself. But, of course, all taxation is a form of socialism. So your indictment of space exploration is not a unique criticism. Same goes for research funding to the National Science Foundation, the National Institutes of Health, and the Centers for Disease Control. By that measure, the National Park Service, the Smithsonian Institution, the National Endowment for the Arts, and the public schools system are also socialist programs. As are the military (since we no longer sell war bonds) and law enforcement agencies. And let's not forget the Environmental Protection Agency, Veterans Benefits, the interstate highway system, and airport infrastructures.

At the end of the day, America is a portfolio of spending that captures and expresses the values of its residents, via its lawmakers.

What a fascinating experiment it would be if we all paid taxes by checking boxes on a form (which is basically what Congress does for every year's budget cycle, but they do it with the populace in mind, not the individual). And suppose it was not a democracy, where majority rules, and you leave the box for NASA unchecked. What happens next? Do anti-tax compatriots come to your home and remove everything from it that was informed, inspired, influenced, invented, or enabled by the space program?

That would make an interesting reality show:

- Gone are the integrated circuits of the electronics you use.
- Gone is the Weather Channel from your cable service.
- Gone is your awareness of satellite maps (from any news source) that track developing storms, hurricanes, and tornadoes.
- Gone is the GPS system in your car (time to buy paper maps again, if you can find someone who sells them).
- Gone are all battery powered hand tools from your garage.
- Gone are (some) of your loved ones from breast cancer because the space-imaging algorithm that detects cancer cells earlier than ever before would not have been used on them.
- Gone is the collision warning system in your car, or in the car you might soon buy.
- Gone is the information to you about asteroid Apophis, which is headed toward Earth right now, for a close approach on April 13, 2036.

- Gone are all satellite news broadcasts from Europe and elsewhere around the world to your television.
- Gone is your awareness that something bad happened on Venus (leaving it with a 900 degree F. runaway greenhouse effect) and on Mars (which once had running water but is now bone-dry and frigid) which themselves have informed the international awareness of global warming.
- Gone are aerodynamic efficiencies of the wings on your airplane (remember the first A in NASA stands for aeronautics).
- Gone is your access to Google Maps.

On a more philosophical level . . .

Gone is your knowledge of our place in the universe—the only human pursuit that has transcended culture, region, and time. All supplied by the Hubble Space Telescope, the Martian rovers, and the countless other spacecraft, with and without humans, that have left Earth to explore.

Others who do support space exploration would have access to these things. But not you. All because you would not annually check a box that allocated 6/10 of a penny of your tax dollar. That is the entire allocation to NASA. That is the cost of your access to the universe that you so willingly decline.

How much is the universe worth to you?

Sincerely,
Neil deGrasse Tyson

Feed Christians to the Lions?

In December 2005, Robert, a devout Christian, took issue with Darwinian evolution in particular, and the findings of science in general, wherever and whenever they conflict with scripture. He was sure that scientists see religious people as their enemy and that if scientists were in charge, we might feed them all to the lions. A point that I think was half-serious. My reply was a long and sustained response to each point that he made, in turn.

Dear Robert,
Nothing in biology makes sense except in the light of evolution.* In modern times, with a thriving growth industry of biotech companies and other business sectors that research the future of our species in relation to all others, if you say, "I don't believe the theory of evolution, I think we were all specially created," you must brace for the consequences of that view to your own employability.

If you don't want to become a scientist, then maybe it doesn't matter. There are plenty of professions that do not involve scientists. But as I said, the emergent economies are going to be scientifically and technologically driven, with biotech front and center. If you're coming in saying that there was Adam and Eve, you're not going to get past the front door.

* A phrase first expressed by the Ukrainian-born American geneticist Theodosius Dobzhansky (1900–1975) who also happened to be devout Eastern Orthodox Christian.

There are simply fewer job options in industries that require a working knowledge of biology, chemistry, physics, geology, astrophysics, in order to make discoveries. I see no reason why other jobs would not be available to you. But the more important point is that current trends indicate that health science fields may be the future of economic growth, so you would not participate in this economic fertility.

In a nation (America) where Pew* polls show that 50% of people believe that there was an Adam and an Eve as the original God-created humans, and that 90% of people believe in a personal God who listens to prayers, you have no foundation to suggest that popular culture will stigmatize you.

You are correct to presume that I celebrate tolerance and diversity, especially of cultures, languages, traditions, etc. But you want this outlook extended to the subset of Christians who take the word of the Bible as literal truth? At any level where claims can be tested—no matter who is making the claim—the issue is no longer about tolerance, it's about objective truths.

For example, nowhere in the Bible is Earth described as a three-dimensional object. All references to it are flat. And until the 15th century, so too were all maps of the world, informed by scriptures. We can celebrate the cultural history of this notion, but it's objectively false. Same holds for the value of Pi. In the

* David Masci, "Religion and Science in the United States: Scientists and Belief," Pew Research Center, Nov. 5, 2009, http://www.pewforum.org/2009/11/05/scientists-and-belief (accessed Jan. 2019).

Bible, a passage (I Kings VII) can only be true if Pi exactly equals 3.0. But we know better than this (so too did the earlier Babylonians, who calculated that Pi was a number between 3 and 4). But just because the Bible says Pi = 3, does not mean Pi = 3. The statement is objectively wrong, and is therefore not a matter of opinion. The fact that the people who wrote the Bible made Pi = 3 and Earth a flat disc is of some historical interest and worthy of study in history class, philosophy class, or religion class. But it has no place in the field of science, whose goal is to find truths of the universe that sit independently of opinion.

Neither I, nor anyone I know, have any intent in feeding Christians to the lions, only in keeping religion out of the science classroom. By the way, there is no tradition of scientists knocking down the doors of Sunday School telling the preachers what to teach. Scientists don't picket outside of churches or shoot people who enter them. Scientists have no tradition of heckling preachers during sermons. And, by the way, nearly half of all scientists (in the West) are religious and pray to a personal god.

You also "accuse" me of being religious—that I follow the religion of science and humanism. Actually I am agnostic.* But I guess I do not know how you are invoking the word. Allow me to find a definition

* *a-gnostic*: A term coined by the nineteenth-century naturalist Thomas Henry Huxley in reference to a person who claims neither faith in, nor disbelief in, God. Today, it references a person who allows the possibility of a God, but remains skeptical.

because I hate arguing semantics. I would rather argue ideas.

Here is one from Webster's:

> *Religion* (noun): the belief in and
> worship of a super-human controlling
> power, esp. a personal God or gods.

Based on that definition, if you think I am religious, then I'm not sure you know what science is or how and why it works, and which has found success precisely because of its empirical approach to nature rather than a spiritual one.

You declare that neither of us can prove our religious beliefs. But I can (and do) know the shape of the Earth, Moon, stars, and universe, the origin of the chemical elements, the age of the Earth and the universe, the extinction episodes of the fossil record, the impact of asteroids on Earth, the genetic commonality among all life on Earth, the genetic proximity of chimpanzees and humans, and countless other objective truths about the world. So your assertion is false and shows a lack of education about the process of science and the nature of discovery. When this happens, it's usually not the person's fault, and is traceable to educators in your life who did not spend enough time training you *how* to think, as distinct from *what* to think.

As for education, I think there ought to be a class in public schools on religion. It occupies an undeniably significant role in civilization. Commensurate with my earlier interest in diversity, the religion class should cover all the world's faith-based philosophies and belief systems. I think, historically, such a class was omitted because religions themselves are not fans of tolerating other religions. And so religious exposure was left to Saturday or Sunday and as a family affair. Which, in retrospect, is probably for the better.

You called me a fraud, which compelled me to look up that word too. Here is what I found:

> *Fraud* (noun): wrongful or criminal deception intended to result in financial or personal gain.

I don't see how that word applies in this case. I think I have been quite open, straight, and honest with you. Yet you took this as a personal attack on Christianity. When in fact it's simply an observation on scientific illiteracy in America.

Thank you again for your interest. I mean that sincerely, as I hope is evidenced by the time I spent in reply.

Neil deGrasse Tyson

Chapter 5

Science Denial

Some people don't like scientists. Some people think that science is a nefarious, political force on society. Some people think that science is over-valued and rampant with smug researchers. Some people are simply in search of what is true. In this chapter, I engage them all.

Middle School Skepticism

Sunday, April 1, 2007
Dear Dr. Tyson,
I am a middle school student and I came across a video that featured scientists who were skeptical about global warming.

My main question to you is: do you think that human-caused global warming is real and worthy of further exploration?

Thank you so much for your time,
Ray Batra

Dear Ray,
There will always be dissenting scientists for any new research findings. What matters most is the peer-

reviewed, published data and the trends in research they indicate. I'm familiar with the video you reference. It interviews the half-dozen or so leading anti–anthropogenic global warming scientists, and a bunch of other non-scientists, like politicians.

Nothing in principle wrong with dissenting views. But since global warming has political and economic ramifications, money flows easily to create videos that feature this subset of scientists. I reviewed the published literature from one of them. He is surely a climate scientist, but not in the field of climate change. His anti–climate change publications are predominantly Op-Eds for newspapers, and other non-peer-reviewed publications.

Compare this with the publication record of NASA's James Hansen and there is no contest who is closer to the problem. Couple that with the huge body of peer-reviewed literature, authored by climate change scientists—and not just climate researchers—and there is no meaningful case to be made. You can find a few dissenters, but they have no data, or they are selective about the data they cite.

Scientists are all human, complete with human frailties and biases and susceptibilities. That's why trends in the data remain the primary commodity of what is true in science, and not the impassioned testimonies of scientists themselves.

Sincerely,
Neil deGrasse Tyson

More Harm Than Good?

Thursday, March 19, 2009

Mr. Tyson,

Has the pursuit of scientific knowledge led to more harm or good for life on this planet?

I just want to clarify and make it clear that I don't intend to attack you or the pursuit of scientific knowledge. I am a supporter of science and do believe that today it does more to help us than hurt.

My question is more to the fundamental issue of whether we, as humans, have ultimately caused possibly fatal damage to our planet through our activities that we must acknowledge fall under the category of scientific pursuit. Gunpowder, coal-fired power, the internal combustion engine, nuclear weapons—these are scientific contributions to life on Earth.

I do believe that these may in fact have been inevitable innovations, in some way, once we left the savannah and began developing technology that enabled us to survive outside our ecological niche.

But, as you are a deeply thoughtful and brilliant human, more even so than just a scientist, I wanted to ask you if you'd ever considered that question: If we could take it all back, wouldn't it really be better for this planet? Not just us, humans, but all life?

Anyways, thanks for your excellent work in spreading the word of science in the modern world. Whatever we may have done before, we certainly need science now!

Best Regards,
Dakkan Abbe

Dear Mr. Abbe,

Thanks for your letter.

I think a list of what is good about science will far outweigh the list of what is bad about science. But what actually matters is that science is not inherently good or bad. It's just a base of knowledge about how the natural world works. It's the engineering applications of science that take on patinas of good or evil. And since no country with any real power has ever elected a scientist or an engineer as its leader, the people who wield resources to fund such good and evil are politicians. So your question could just as easily have been reworded with the word politics swapped in for science.

Control of nature is not unique to humans. Beavers wreak havoc on their environment. We have revisionist comments regarding what they do: "Their dams create a habitat for all manner of wildlife," when, in fact, their dams completely change the local ecology. Swarming locusts and cicadas also create imbalances in their habitats. But the worst of them all? Four billion years ago, cyanobacteria transformed Earth's atmosphere by making O_2 in the greatest ecological disruption in the history of life on Earth, killing all surface-living anaerobic bacteria.

Human-induced global climate change is (at the moment) not unstoppable. And, of course, the solution will come from science and technology—via enlightened leadership. Just as the problem arose from science and technology—via shortsighted leadership. But this cycle is nothing new.

We solved the food shortage problem in the world*—which was as big a fear at the end of the 19th century as global warming has been in the 21st century. We have also (in America) made great progress on the pollution problem, after it was identified and articulated in the 1970s. The Environmental Protection Agency (EPA) was formed to oversee this effort, and now the rivers, lands, and air in America are cleaner than at any time since the industrial revolution began.

Many worried that science applied to farming and animal husbandry might eviscerate foods of their nutrients or flavor. Some of this indeed happened. So today (in America, but especially in Europe) there's a huge and successful movement toward local produce and organic farming.

I therefore retain a confidence that you do not have—that science has the power to solve the problems it *occasionally* creates, provided the political and cultural will is there to enable it.

And I maintain that without the progress of science, I, today, would be someone's slave, and half the world would not have not survived past age five. Not only that, 70% of who did survive would be hard at work on farms, barely making enough food for the growing population.

* Of course millions of people, mostly children, starve each year. But this is traceable to bad politics and broken distribution channels, not to a shortage of food in the world.

But I nonetheless thank you for your question, your interest, and your kind remarks about my work.

Sincerely,
Neil deGrasse Tyson

Evolution vs. Creationism

Sunday, August 3, 2008
Dear Dr. deGrasse Tyson,
I have been seeing quite a bit of conflict regarding teaching evolution vs. creationism. If I've read correctly, you believe in evolution (which I do, too), but does this mean that you don't believe in "God" or a higher power?

I have become so confused about what I believe. I was raised Catholic all my life (went to an all-girls Franciscan high school and Marquette University, a Jesuit university), but I have very serious doubts about a higher power. We are such a speck in the grand scheme of things . . . actually less than a speck. So I just wonder what your feelings are. I hope I did not ask a taboo question. If I did, I'm truly sorry. If not, I look forward to your response.

Thank you, Dr. deGrasse Tyson.

Sincerely,
Jackie Schwab

Dear Ms. Schwab,
Thanks for candidly sharing your angst about a higher power.

A few points . . .

The theory of evolution is not something to "believe in." Science follows evidence. And when strong evidence supports an idea, the concept of belief, when invoked the way religious people use the word, is unnecessary. In other words, established science is not an ensemble of beliefs, it's a system of ideas supported by verifiable evidence.

You did not ask if I believe in the sunrise. Or if I believe the sky is blue. Or if I believe Earth has a Moon. These are non-controversial truths about the physical world for which the word "believe" has no place. Evolution by natural selection is a non-controversial tenet of modern biology. It's not controversial among biologists, that is. Among religious fundamentalists, biological evolution does not square with their faith-based belief system, in which they invoke the Bible as an inerrant understanding of the physical world.

This leads to claims, for example, that Earth is no more than 10,000 years old. And that there was a literal flood in which all of Earth was covered in water. No evidence supports this. Not only that, all evidence argues against it. And so one is left "believing in" stories that are demonstrably false.

Thanks again for your interest and your questions.

Neil deGrasse Tyson

Qur'anic Verses

On Wednesday, June 3, 2009, Tahmid Rahim, * *a Muslim, respectfully inquired why, in my various appearances in science documentaries and elsewhere, science in the Qur'an was never mentioned. He declared that the Qur'an contains many verses that reference specific discoveries of modern astrophysics, from relativity to the expanding universe. For a book written by Muhammad 1,400 years ago, this would be extraordinary, if true.*

Hello Tahmid Rahim,
Thanks for your note.

A great challenge of revealed truths by divine prophets is that nobody has ever made a successful prediction of previously unknown objects or phenomena based on the contents of any religious texts. What typically happens is that devout people learn what scientists have discovered about the natural world and then go back into their religious texts in search of passages that hint at what is already known. But since the extracted information comes after-the-fact, it's not useful to the advance of science. What you need to do, if you are convinced of Qur'anic foresight and inerrancy, is come up with *predictions* about the natural world derived from Qur'anic verse that will stimulate research. If any of it comes true (it would be a first time for such a thing, by the way)

* Name changed.

then scientists would be mining the Qur'an for its insights daily.

This has never happened—with any religious text—which is why they have no place in the science classroom. Occasionally, when strongly religious people feel that science conflicts with their religious texts, they fight the concepts, declaring something is wrong with the science.

Come up with a list of predictions of unknown phenomena, derived from the Qur'an, and I will be happy to offer comment. Otherwise, science and religion don't have much to say to each other.

Sincerely,
Neil deGrasse Tyson

Evidence for God

In a long exchange from 2008, Andrew McLemore expressed his enthusiasm for science as a tool to peer into God's cosmic handiwork. But he wondered what level evidence might convince a skeptic that there's a better-than-even chance that God exists.

Dear Andrew,
I think often about what would constitute evidence for God. How about if, after adjusting for income and access to health care, pious people all lived longer than non-pious people? How about when a plane crashes, only the pious people survive? How about Jesus comes when people say he will come? (His sec-

ond coming has been predicted by Christians for hundreds of occasions that have already passed, spanning the last 2000 years.)

How about people pray for peace, and then all wars in the world stop permanently? How about good things happen exclusively to good people and bad things happen exclusively to bad people? How about an earthquake strikes Lisbon on All Saints Day, while everyone is in Church, as it did in 1755, and it kills only people who are not in Church, rather than the tens of thousands of people who were, as what actually happened that fateful morning.

These events would trigger serious (scientific) conversation about the existence of God and how he treats people who worship him versus those who do not.

Sincerely,
Neil deGrasse Tyson

Where Is the Proof?

In June 2008, Roger argued heavily against the discoveries of science that conflict with biblical statements about evolution and the age of the physical universe. He even called me an arrogant liar. Based on that name-calling alone, our exchange might have landed in the Hate Mail bin of this volume, but he fundamentally disputed major discoveries of modern science, which instead lands it here, in the Science Denial chapter.

Roger,

You doubt all dating methods that extend timelines of the world beyond that of recorded history. Whatever is the source of your denial does not hold your intellectual enlightenment as its priority.

Measurements obtained by separate groups using distinctly different methods applying different principles of investigations have shown that:

- the age of meteorites is 4.55 billion years +/-0.01
- the age of moon rocks is 4.55 billion years +/-0.01
- the age of the Sun is 4.5 billion years +/-0.1
- the age of the oldest crust on Earth, a planet that recycles its crust in and out of volcanoes, is 4 billion years +/-0.01

Carbon-14 isotope dating is effective for no more than a few tens of thousands of years, and is useful primarily on material that was once alive. So it's widely used to date cave artifacts from the stone-age times. But the isotopes of various other elements on the periodic table are useful for time intervals of millions, tens of millions, hundreds of millions, and even billions of years.

After it forms, one can measure what fraction of a radioactive element decays into another element. These are called daughter elements. The bigger the fraction of daughter elements in the sample, the older the sample is. It's that simple. Some elements decay much more slowly than others, making them useful for reaching and dating longer periods of time.

We determine the age of the Sun from calculations based on its mass and the rate it consumes energy, two easily measured quantities. This requires knowing that the Sun produces energy by thermonuclear fusion of hydrogen into helium.

None of these are controversial results. We are all on to the next problems. If these results sit uneasily to some people, I have found that it almost always derives from conflict with a pre-existing expectation of how the universe must be.

You further wonder how, if humans evolved from apes, why apes stop evolving. Natural selection drives evolution. And evolution is occurring all around us. All the time. It's best seen among species that have fast reproductive cycles, where variations can be selected and identified on time scales short compared with a human life. The bacterial branch of the tree of life is huge—much larger in variation than what is found among vertebrates, for example. Among bacteria as well as viruses, we see speciation all the time. Some of the more visible ones are Swine Flu, AIDS, and Legionnaire's disease. These afflictions did not exist in nature until they mutated from previous forms, becoming new species, enabling them to infect life not previously accessible to them.

Not all species are evolving at all times. For example, the Coelacanth (pronounced See-luh-kanth) is an extremely successful bottom-feeding fish that has not changed substantially for the past 360 million years. Horseshoe crabs go back even further—450 million years. If you are a successful species, there is no driver to influence change. Meanwhile, mammals have changed dramatically over the past 65 million years. When I say dramatically, I mean visually, not biologically. We share more than 90% identical DNA with all mammals, even mice.

Among mammals in the tree of life is the branch called primates, such as lemurs, monkeys, and the great apes, including humans. It's commonly thought that humans evolved from monkeys. But this is not true. We all have a common ancestor. The ape that is closest to us is the chimpanzee. In other words, chimps and humans have a relatively recent common ancestor.

As you would expect from this information, we are, in fact, genetically closer to chimpanzees than to any other animal in the world. Contrary to your assertions that chimps and humans are completely different from each other, chimps and humans have every muscle and every bone in common. Chimps and humans even have the same facial expressions. But most importantly, we have only trifling differences in our DNA. In fact, genetically, we and the chimps are closer to each other than either of us are to the "Old World" monkeys of Africa.

I call this to your attention because your two emails to me were not in the form of questions. They were declarations of information, as though drawn from some source that you trust. But, like I said, that source does not have your science literacy or intellectual enlightenment as a priority.

Sincerely,
Neil deGrasse Tyson

Chapter 6

Philosophy

Sometimes you just have to ask a deep question.

Alien Homicide

In February 2007, Michael Cuellar inquired about the legality and morality of killing a visiting space alien who may be more intelligent than we are. Or does might make right?

Hello Mr. Cuellar,
I do not claim to be an expert on morality, but I am happy to offer an opinion and perspective on your queries. Yes, it would be morally wrong, unless we were starving with no other source of food and their flesh were digestible to our stomachs.

I would think it's morally wrong to harm anything, regardless of its measure of intelligence, without reasons of promoting your own survival or the survival of your kin. I cannot imagine who would think that it's not wrong to do so. There is a growing literature of Space Law that addresses the meaning of murder if you kill a visiting space alien, whose civil rights are not protected under any constitution in the world.

Furthermore, "might makes right" is not the same thing as "might makes morality."

No doubt we'd have a hard time killing a more intelligent species than we. If we assume they are more intelligent compared with us than, say, we are compared with chimps, then they would not fear us any more than we fear an uprising in the monkey forest.

To keep our identity secret will be very hard, now that our radio bubble is more than 70 light-years out there, and expanding.

Thank you for your interest.

Sincerely,
Neil deGrasse Tyson

Truth or Meaning?

Tuesday, September 20, 2005
Dr. Tyson,
I am a high school science teacher (astronomy and physics) and a tremendous fan of your work.

I am also currently pursuing my doctorate in Educational Psychology (University of Illinois-Chicago). I am engaged in a lively debate this semester as to the role of science in research. Distilled, the question comes down to, "Is science concerned with truth or understanding/meaning?" I would value your opinion on this matter.

Best Wishes and Clear Skies,
Kevin Murphy

Dear Mr. Murphy,
Thanks for your note.

I have never been a big fan of philosophy as applied to the physical sciences in the 20th (+21st) century. I have found common arguments to be based more on word usage and word meaning rather than on ideas, and so have found the discussions to be largely useless to the progress of science—a domain where ideas matter more than words.

So I refuse to enter arguments over words. I would rather state what science does, and leave you free to attach whatever words you wish to the enterprise. If we agree on the word, then fine. If we do not agree on the word, the expressed idea remains unaffected.

That being said, science seeks all three (truth, understanding, meaning), but primarily concerns itself with obtaining sufficient knowledge of how the universe works to be able to make testable predictions about its past and future behavior. Computer simulations can substitute for past and future behavior when sensible.

If we predict with accuracy and precision the behavior of nature, then we are satisfied with our work and move on to the next problem. I would say that the major equations of modern physics represent cosmic truths. As do the major ideas of how the universe works—quantum theory, relativity theory, evolutionary theory, thermodynamic theory, etc. These truths provide understanding for the behavior of things and the existence of phenomena.

The word "meaning" is more often than not taken to be personal. The way people often use the word specifically excludes science and its methods and tools. But one can imagine a new kind of philosophy where science is brought to bear on social/political/cultural issues. For example, if you a-priori assert that human life is sacred, then decisions related to saving and preserving it become simple matters of reason. If vacation and home life adds meaning to peoples' lives, then you would use the methods and tools of science to help you make decisions that maximize this feature of life. Right now, these issues and others are inefficiently argued by politicians, religious leaders, and lawyers.

Good luck with your studies. And thanks again for your comments and your interest in my views.

Neil deGrasse Tyson

How?

Wednesday, March 16, 2005

Dr. Tyson,

I had the pleasure of attending your lecture last night with my two colleagues. The topic of the convergence of Science & Religion has fascinated me for years as a scientist and as a religious person.

I wholeheartedly agree with your conclusion that using religion to explain frontiers of science is inherently short-sighted. Over the past several years, I have read books where a common theme in all of them is a need to separate science's goals (explaining "how") and religion's goals (explaining "why"). When either of the two attempt to explain the others' main goal, they necessarily fail.

One philosophical note: it is my personal feeling (regret), that science is converging with religion because it is becoming a religion. The absolute faith that science can explain everything (i.e. "Scientism") is unfortunately embraced by those who do not see they are creating this new religion. In politics, the principal of Secularism closely parallels this.

Thomas E. Downs

Dear Mr. Downs,
This distinction between "how" and "why," while resonant with emergent philosophies, is not entirely clean. The following are just a few (pure) "why" questions for which faith-based inquiry is powerless to answer, beyond the catch-all response of "Because God made it that way":

- Why is the sky blue?
- Why does the moon always show the same face to Earth?
- Why does Venus go through phases like the Moon?
- Why does the Sun have spots?
- Why do hurricanes in the northern hemisphere rotate counterclockwise?
- Why is August hotter than June, when the Sun's rays on Earth are more direct in June?

Consider further that I know of no book of "why" questions, answered unambiguously by faith-based religious philosophy—that is, "why" questions answered in a way that everyone can agree. If faith is a personal construct, then there can be no agreed-upon book of answers.

Active scientists do not run around declaring that science can explain everything. For example, no one of us claims that science can explain love or hate or beauty or valor or cowardice. But as science advances, these notions may indeed come into the experimental purview of science, just as so many previously intractable subjects have done in the past. This is not an absolute faith as you describe it, it is a prevailing confidence based on past performance of the methods and tools of science.

Faith, as the term is commonly used, requires no experimental evidence to hold a belief. So to declare that science is becoming an absolute, faith based religion generalizes science in a way that is simply untrue among the actual practitioners of the field. I find this argument used principally when the arguer hears the word "faith" used pejoratively, and so invokes it as an attack on science so that science does not retain that philosophical advantage over religion.

Thomas E. Downs continued . . .

Finally, don't get me wrong—as a scientist myself, I realize that "active scientists do not run around declaring that science can explain everything." However, public opinion (mostly a result of ignorance) does lead to this conclusion when science is presented as an attack on religion.

You know as well as I that the intention of most scientists is not one of anti-religion, but there are a few out there who revel in it and the backlash it creates.

One quick clarification. For sake of space, from Merriam-Webster.

- *Why: For what cause, reason, or purpose.*
- *How: In what manner or way; to what degree or extent.*

I followed . . .

To base a philosophy on a dictionary distinction between two related words is shaky ground. All too often, modern philosophical arguments are traceable to disagreements in the definitions of words rather than in the analysis of ideas themselves.

"Why is the sky blue?" is a question that seeks a cause, just as required of the definition above. I suppose we could re-word the question to use "how," but the sentence would be awkward and not represent the way anyone thinks about the problem in real life: "How does white light coming from the Sun turn blue as it passes through the atmosphere?"

From the other direction, one could ask: "Why am I here?" A common, simple construction. But I assert that this can be turned into "how" questions with similar effort: "How did inanimate matter assemble to become animate? How did animate matter evolve to become Homo sapiens? How did the run of Homo sapiens lead to me, here and now?"

I think the real issue here is not "how" or "why," but the questions themselves, without regard to the definition of the first word in the sentence. We can make a book of questions about the world that are answered by science. And this book continues to grow exponentially, with a doubling time of 15 years (based on peer-reviewed research publication rates in all sciences).

Is there a book of questions with answers provided by spiritual inquiry? (Of course, religion has been at it for many thousands of years.) If such a book exists, how big is it? Is this book growing? Does the book distinguish itself from other written works that probe the human condition, like the complete works of Shakespeare?

So while I will not declare that science can currently answer all questions, the trend is quite impressive, especially when compared with religion, which spent most of its history explaining things (answering whole categories of questions) via divine forces that, in fact, had natural explanations, such as disease, hurricanes, planetary orbits, etc. Note that natural disasters are still referred to in many insurance contracts as "Acts of God."

I note further that religion-bashing occurs primarily among atheists and not by scientists (while there is, of course, some overlap, the loudest atheists are generally not scientists). And yet, my read of modern culture tells me that attacks by religion on science are vastly more common than the reverse, contrary to your statement. Recently in Georgia, the school board wanted to put a disclaimer sticker on biology textbooks. But you do not find scientists asking for disclaimers to be affixed to Bibles in church.

The most visible anti-religion scientist I know is the physicist Steven Weinberg. And he is far outnumbered by pro-spiritual scientist/writers such as Paul Davies, Robert Jastrow, and John Polkinghorne.

And may I remind you that in the famous Scopes "monkey" trial,* the science teacher lost the case.

Neil deGrasse Tyson

Why?

Circa 2009
Via Facebook

May I ask you two quick questions?

1. *Are you a Sam Cooke fan?*

2. *What's your honest opinion as to why we are here?*

Jason Harris

Dear Jason,

1. I'm not any more of a Sam Cooke fan than I am of other crooners in the era.

* The Scopes trial, *The State of Tennessee v. John Thomas Scopes*, July 1925. Substitute high school teacher John T. Scopes was accused of violating Tennessee law by teaching evolution in the classroom.

2. I never think much about "why." Why implies purpose set by external forces. I have always felt that purpose is not defined outside of ourselves, but from deep within.My purpose in life to lessen the suffering of others; advance our understanding of the universe; and enlighten others along the way.

Neil

Yin and Yang

Circa 2009
Via Facebook

Neil,
Everything I have learned and observed in this world and in this universe seems to fit into the idea of Yin & Yang. Everything flows in cycles from biological and physical realities to ideologies, to presidents. However, as I understand it, the prevailing astrophysical view of the end of the universe is one that fits the entropy law: greater and greater disorder until everything is basically as spread out as it can get. This seems to me to be the only example of Yin & Yang being violated.

I understand that nothing has been shown to violate the entropy law. However, the principle of yin and yang seems to work inside the entropy law. Is this the case in your mind? Are there any oscillatory theories of the universe that would allow me to reconcile these beliefs? What are your thoughts on the matter?

Reid Tice

Dear Reid,
Yin & Yang offers no predictive value unless you can invoke its principles to say when and where something will cycle back. Apart from this, my read of Yin and Yang is not that things cycle but that things are in balance—with opposite forms, themes, ideas, in mutual, though beneficial, tension.

Furthermore, many things have not cycled back—and will not likely ever cycle back. There's no longer any state-sanctioned slavery. Kings have lost most of the power they once wielded in war, culture, and politics.

Mars was once an oasis of running water. It's bone dry today. And there is no evidence that it will ever return to its previous state. Same for Venus, on which there's a runaway greenhouse effect, leaving the surface at 900 degrees Fahrenheit.

We are living longer than ever before. This advance of technology, and its role in our lives, is not a reversible trend. So you cannot ignore all that does not cycle, select what does, and declare Yin and Yang to be an operational principle of our universe.

Neil

I Think, Therefore I Doubt

Wednesday, May 20, 2009
Dear Mr. Tyson,
I am somewhat torn. I can't get near philosophy without being repelled by its un-scientific musing and empty wordiness. I simply do not understand how one can be so confident that

their particular explanation for the universe, or consciousness, or the meaning of knowledge is anywhere near correct without the necessary experimentation and peer review. Can this field be taken seriously when, to debate another person's point of view, one must only summon up their own, equally unfounded, ideas?

But, many of these philosophers were also very intelligent men. Some were even scientists themselves. Certainly if these intelligent men muse to themselves, it could have some merit. This leads to my dilemma: I do not know how to reconcile the fields of philosophy and the fields of science, except to say that philosophy simply muses about things science hasn't explained yet. To me it is a more relaxed and vague form of theology.

So I ask you: What do you think of philosophy's role in the explanation of the workings of the mind and universe, and in the field of science?

Thank you very much for your time.

Respectfully,
Daniel Narciso

Dear Mr. Narciso,

My sentiments largely align with yours. I have yet to see a philosopher, formally trained in the 20th century and onward (via a university Philosophy Department) make any material advances in our understanding of the natural world. They typically carry a level of confidence in their knowledge that is unwarranted by data and observations of the physical universe. Philosophers have no laboratories. No telescopes. No microscopes. They have their brains and armchairs, and falsely believe that this is sufficient to gain insight to the operations of nature.

I have no comment on other branches of philosophy: ethics, religious philosophy, political philosophy, etc. I lament the loss of useful philosophers that predated modern physics—Immanuel Kant, David Hume, Kurt Gödel, Bertrand Russell, Ernst Mach. Not coincidentally, the transition to uselessness began when our experiments revealed aspects of the universe that no longer followed what anyone would call common sense. The tenets of relativity and quantum mechanics, for example.

The day a philosopher's conversation on the "meaning of meaning" offers useful insight to the next cosmic discovery, I will be happy to revise my views.

Best to you,
Neil deGrasse Tyson

Express Yourself

Undated Communication—circa 2014
Via US Postal Service

To: Neil deGrasse Tyson
I have seen you on the History Channel, Discovery Channel, I bought and read your books. And I have heard you on the overnight radio show "Coast to Coast."

One thing has remained constant throughout all these media forms; and that is how you communicate; the style that you express and transfer information, which leads to my question.

Who, what, where, and how have you learned to communicate so effectively. I have a lot of information (in my head) that I have trouble expressing effectively. You seem to the able to anticipate what questions a person may be thinking while they are reading or listening—and then answer them in the next sentence or paragraph. I wish to learn how to be so foresighted.

I have included a return envelope for your convenience. The Number and Unit designation on the envelope is my address at the Texas prison where I am housed.

Thank you,
David Swaim #1436288
Iowa Park, Texas

Dear Mr. Swaim,
Thanks for your kind words about my efforts to communicate.

My educational philosophy is quite simple. Think of a professor facing away from you, droning on while writing on the board in the front of class. As a student, especially in college, it's your responsibility to learn the material. You are paying for the education. So your learning skills will, in many cases, need to compensate for the absence of clarity or enthusiasm in the professor's delivery. That's lecturing.

Now think of a professor who faces you from the front of the room; who makes eye contact with the audience; who has invested time and energy thinking about how you think; who pays attention to your attention span; who is aware of what words you know and what words or concepts confuse you; who knows the demographics of the audience—age, gender, nationality, ethnicity, political leanings, cultural leanings, propensity to laugh, to cry; who carries some pop-culture fluency, for easy reference and analogy, but only when teaching the subject can be assisted by such references. That person is not lecturing to you. That person has opened conduits tailored to that audience in that moment, and at that time. That's communicating.

It's a way to see and feel what someone is thinking, allowing you to serve their curiosity on the spot.

Also, most things I've written for publication passed through at least two editors—English majors in college who care about the language. In at least one of my books I thanked my editor for helping me "to say what I mean and mean what I say."

So there are no shortcuts. But you'll know when you have perfected the task, that's when under-informed people come up to you and say, "you're a natural at this."

Sincerely,
Neil deGrasse Tyson

III.

Pathos

A plaintive appeal to emotions that already reside within us.

Chapter 7

Life and Death

Life is never easy. Death is even harder.

Remembering Holbrooke

New York Times headline:

> "Richard C. Holbrooke, 1941–2010: Strong American Voice in Diplomacy and Crisis"

Thursday, December 16, 2010
New York Times

To the Editor:
When I gave Ambassador Richard Holbrooke a personal tour of the newly opened Rose Center for Earth and Space and Hayden Planetarium in 2000, I could not help notice how fluent he was in the depth and breadth of his cosmic curiosity.

True science literacy is less about what you know and more about how your brain is wired for asking questions. Later in the tour he confessed that, as an undergraduate at Brown, he studied physics before switching to politics.

I could not resist asking him whether that exposure to physics made a difference in his career as a diplomat, especially in tense, war-torn areas of the world that are resistant to negotiated peace settlements.

He answered emphatically "yes," citing the physics-inspired approach of sifting for the fundamental drivers of a cause or phenomenon—stripped of all ornament. To get there, one must assess how and when to ignore the surrounding details, which can give the illusion of importance, yet in the end, are often irrelevant distractions to solutions of otherwise intractable problems.

Mr. Holbrooke's career was a living endorsement for more scientifically literate peace negotiators in the world.

Neil deGrasse Tyson
New York City

Dead Man Talking

Wednesday, March 27, 2019
Dear Cousin Neil,
The day after my father died, I went to the funeral home to view his body. He'd been sick for nearly a decade following a series of debilitating strokes and his death, while painful, was expected.

Walking into the funeral home I could just barely see his body stretched out on a table in front. I gathered my courage and knew the time had come for me to say my goodbyes. Just then, I heard an old familiar voice say to me "What the f%#k are you doing kid? Get out of here!"

I stopped dead in my tracks and spun around only to find no one there.

I knew that voice, it was one I hadn't heard in ten years. My dad's stroke had left his voice forever changed; yet I knew to the core of my being that the voice I heard belonged to my father.

Hearing him use the words 'kid' and 'f%#k' also confirmed it was definitely him. He always called me kid, and to him, f%#k was just an adjective.

Without thinking, I said (out-loud) "I'm here to see you." He said, "I'm not there!" I started to leave and stopped in my tracks, turned around and said, "No! I came to see you and I'm going to see you!" He said, "Fine, go look."

As I walked up to the body I wasn't sad anymore. As I looked down at him, his body looked waxy and his face misshapen from the breathing tube he had been on, I heard him say, "See? I told you I'm not here."

Happier and more peaceful than I had been only minutes before, I literally skipped out of the funeral home. Years later, it still feels real but logically doesn't make sense.

What do you think really happened?

Seanlai Cochrane
Delray Beach, FL

Dear Seanlai,
Either my first cousin (your dead father) was actually talking to you, or you acoustically hallucinated his voice. While the latter is far more likely to be the case, allow me to suggest an experiment for you to conduct if anything like this happens to you again.

Next time somebody dead talks to you, try to have

a more informative conversation. Try to glean information about the great beyond. Be curious. Ask good questions. Here are a few that come to mind.

- Where exactly are you?
- Is anybody else there? If so, who?
- Are you wearing clothes? If so, where did you get them?
- Do you eat food? If yes, then who prepares it?
- Describe what you see around you?
- How old are you? How is your health?
- Is there day and night?
- Do you sleep? Where do you sleep?

If you have an active, creative, imaginative brain, it's entirely possible that the hallucinated voice of your father would offer interesting, plausible answers to each of these questions. So to mitigate that possibility, get someone to write a short phrase on a sheet of paper—like "howdy partner" or "diamonds are forever"—making sure you do not see it. Then hold it up and ask your deceased father to read it. You are now soliciting information that does not reside within your own brain.

If you can demonstrate that the dead person knows things (with accuracy) that you do not know, then you will become famous overnight. If you can't, then chalk up the experience to yet another occasion when our brain misleads, distorts, or mangles objective realities.

Neil

Farewell*

Thursday, December 24, 2009
To all of my Professors & Educators,
It may sadden you, but I hope by the end of this note, it won't.

The simple medical fact is, I'm about done. I've been carrying some nuisance issues for about a year and decided to have them looked into, and the short story is, I've got cancer in so many places I stopped listening to the doctor after the first four. It's terminal, and it's short-term.

The one thing I'm going to have to insist on is that I get no "boo-hoo" email. I consider myself a pretty lucky guy. I got out of corporate life in 1995, and retired for good in 2002, and during this time, I've had a really interesting life. For the last seven years, I've had my time all to myself to study science and math, and help beginners in these studies. I've got a dream telescope and have seen wonders in the night sky that most others will never see for themselves. Through all this, the universe has given me a spiritual awakening that convinces me that life here on Earth is nothing more than a phase. And as if all this isn't enough of a prize, I've been blessed with a "two minute warning" to make this transition as orderly and meaningful as possible (and, by the way, time to gain appreciation of the very many things I've been taking for granted for years).

* An open letter sent to twelve favorite professors from the *Great Courses* video series that this gentleman watched and enjoyed. Mine among them. Note: We corresponded six months earlier, in the Parenting chapter of this collection.

You people have made the last several years of my life rewarding—with an objective, with drive, with a purpose. Very many folks spend the last years of their lives with nothing more than trying to find something to do. I'm a floor above those folks—and it was the discovery of The Teaching Company, astronomy, science and math that drove me to these heights that I would not have found otherwise. No, it wasn't you alone—it was also my research and my book studies that pushed me forward—but it was you, collectively, that provided the propellant.*

If you reply to this email at all, please let it to be to wish me well in the fantastic adventure I'll soon be embarking on. My soul is tough, and I'll see it through.

Good luck and thanks to each and every one of you, and don't ever underestimate your contributions. We'll talk again on the other side.

Regards & Farewell,
MJ "Morg" Staley

Dear Morg,
You surely know by now that a cosmic perspective offers vistas that may serve and sooth your current state of body & mind.

And as the saying goes: We're all going to die. But only a select few of us happen to know when.

Neil

Postscript: Morg Staley died eight months later, in August 2010.

* Now *The Great Courses*, Chantilly, VA.

The Cosmic Perspective

Tuesday, June 19, 2012
Mr. Tyson,
Thank you!

My mother is passing away at this time. I have been by her side as much as I can be. I never had much time with her, as she walked a different road in life; she took my sister's hand and I lost many years.

A few years ago, she asked to live with my wife and me. We never really shared anything or discussed much. But you helped her and me to find a point of discussion. Thank you.

We are born alone, we die alone. It's what we do that makes the only things we take with us.

My greatest thanks belong to you.

Regards,
Robert Clark

Dear Robert,
While you haven't specified, I presume the points of discussion with your mother have been various things I've written or said about the universe. One good thing (among many) about the cosmos is that it belongs to us all. As a consequence, the more you learn, the more ownership you're compelled to take of it.

On my deathbed, one thought I will surely have comes from the evolutionary biologist Richard Dawkins. He notes that we who die are the lucky ones. Most people—most genetic combinations of who could ever exist—will never be born, and so will never have the opportunity to die.

That, and other reflections on our place in the universe, never fail to bring me intellectual enlightenment and spiritual peace when I seek them. I would be honored if you read the ending paragraphs of my essay "The Cosmic Perspective"* to your mother, if there's still time between you. Reproduced below.

Strength to you, peace to your mother,

Neil

> The cosmic perspective flows from fundamental knowledge. But it's more than just what you know. It's also about having the wisdom and insight to apply that knowledge to assessing our place in the universe. And its attributes are clear:
>
> - The cosmic perspective comes from the frontiers of science, yet it's not solely the province of the scientist. The cosmic perspective belongs to everyone.
> - The cosmic perspective is humble.
> - The cosmic perspective is spiritual—even redemptive—but not religious.
> - The cosmic perspective enables us to grasp, in the same thought, the large and the small.
> - The cosmic perspective opens our minds

* Adapted for *Astrophysics for People in a Hurry* (New York: W. W. Norton, 2017), 205–208.

to extraordinary ideas but does not leave them so open that our brains spill out, making us susceptible to believing anything we're told.

- The cosmic perspective opens our eyes to the universe, not as a benevolent cradle designed to nurture life but as a cold, lonely, hazardous place.
- The cosmic perspective shows Earth to be a mote, but a precious mote and, for the moment, the only home we have.
- The cosmic perspective finds beauty in the images of planets, moons, stars, and nebulae but also celebrates the laws of physics that shape them.
- The cosmic perspective enables us to see beyond our circumstances, allowing us to transcend the primal search for food, shelter, and sex.
- The cosmic perspective reminds us that in space, where there is no air, a flag will not wave—an indication that perhaps flag waving and space exploration do not mix.
- The cosmic perspective not only embraces our genetic kinship with all life on Earth but also values our chemical kinship with any yet-to-be discovered life in the universe, as well as our atomic kinship with the universe itself. We are Stardust.

Robert Clark replied . . .

THANK YOU. Your strength has helped me very much and your encouragement to my mother has not gone unnoticed by her. My mother's condition has stabilized, but she is still in an acute wing of the hospital.

It seems that knowing those she admired are pulling for her has encouraged her greatly. I will be sitting with her this weekend and I will again read the whole essay to her. She wanted to hear all your words, she reacted better to your words than words from the Bible others have been reading to her (I hate to put you under such pressure).

Thank you again, and always a student,
Robert Clark

Soul Searching

In July 2007, Jeff Ryan wrote with questions about life after death. Is there a soul or essence of ourselves that transfers, gaining eternal existence? But, most importantly to his curiosity, what does science say about it all?

Dear Mr. Ryan,
The human body contains a measurable quantity of energy, stored chemically (in its fat, and in all other soft tissue) as well as energy that derives from existing at 98.6 degrees, a temperature typically above the ambient air temperature and sustained while alive by the release of stored chemical energy within our bodies. We also harbor trillions of symbiotic and parasitic organisms on your skin and especially in your digestive tract.

When we die, our chemical processes (the metabolism) ceases to function and we immediately begin to lose energy to the air as our body cools. The rest of our body makes tasty food for microorganisms that are already in our body as well as others that are attracted to it, such as fly larvae, worms, etc. Over time, the entire energy content of our body returns to the Earth and the atmosphere from which it came.

If you are cremated, none of this energy is made available to nature even though we had been drawing from nature for our entire life for sources of food. When you are cremated, the stored chemical energy gets released into the atmosphere, heats the air, and then radiates energy into space.

For this reason, it is my strong preference to be buried, completing the cycle of energy that began with my birth.

All of this derives from the chemistry and physics of measurable quantities.

If you believe you have a soul, as several of the world's religions assert, then its existence is faith-based and so you cannot appeal to the methods and tools of science to say what happens to it. Unless, of course, you can make a testable prediction about how to measure the soul. This, in fact, was attempted shortly after the discovery of X-rays. People, eager to prove their faith in a soul, identified dying patients in a hospital and X-rayed them at the moment of death to see if anything rose out of their bodies. They saw nothing.

Sincerely,

Neil deGrasse Tyson

Hurricane Katrina

January 27, 2010
Via Facebook

How come everyone is so quick to want to help people in Haiti but it seems everyone has forgotten about the poor and displaced people in America? Why not donate to a charity to help the U.S. instead? There are still people just as bad off because of Katrina as those in Haiti, but no one cares about them.

Ron Marish

Dear Ron,
Scale matters. About 2,000 people died in New Orleans from the failed levees. Meanwhile, the earthquake death toll in Haiti reached a quarter million—nearly 3% of that nation's population. And so the magnitude of the Haitian earthquake *dwarfs* that of Katrina.

Personally, I draw the line when people step across a homeless human in the street to feed or adopt a stray dog.

Neil deGrasse Tyson

Curing Disease

Randy M. Zeitman was interested in the age-old dilemma of whether intellectually talented people should pursue their own interests or devote that mental power to solving the pressing problems of society. He questioned the value of walking on the Moon or the Hubble Space Telescope if we still haven't cured cancer, or if we still have a problem feeding the world. In October 2004, Mr. Zeitman (politely) challenged me on this tension between doing what you want and doing what is right.

Dear Mr. Zeitman,
Thank you for sharing your comments and critical perspectives. I once felt exactly as you do, but changed my mind after I learned some basic (yet not widely appreciated) facts of life and of society.

You referred to a cure for cancer. Tax money spent on cancer and disease research in America exceeds that which we spend on space by a factor of ten. When you include private/corporate R&D spent on curing disease, the factor rises to one hundred. So it is not as though we are not already investing huge resources into these crucial fields. NASA just happens to be the most visible among targets for your line of argument.

Note that you did not compare the cost of curing cancer with money Americans spend on the Defense Department, or on farm subsidies. Why not? The Department of Defense spends in ten days what NASA spends in a year, not including the cost of veterans' benefits. America spends upwards of $100 billion per year in cash payments to farmers so that they do *not* grow crops. This, itself, is more than six times NASA's annual budget.

But what's more important than any of the above comparisons is the fact that truly innovative solutions to problems come primarily from the cross-pollination of disciplines. And this cross-pollination is entirely unpredictable in its nature and its direction. I give several examples drawn from health, but thousands exist in all fields: A new computer algorithm for image analysis was invented back when the original Hubble Telescope mirror was launched and found to be defective. Until the optics were repaired, applying this algorithm was the best we could do with the fuzzy images. Turns out, the algorithm was shown to be ideal for the early detection of breast cancer, enabling diagnosis long before the trained human eye could determine that cancer was present. No doctor has the background to even think about computer algorithms for this purpose. And the same is true for X-ray machines (invented by physicists exploring the electromagnetic spectrum), the MRI machines (concept discovered by physicists), and ultrasound devices, developed by the military for undersea surveillance.

I will add that my visibility as a Black scientist serves to break stereotypes that, themselves, take an incalculable toll on society, as opportunities and resources are closed off because people in power do not see people of color as having the intellectual mettle to compete in the workplace, in academia, or elsewhere.

So I could not disagree more with your contention. The facts of how society works argue strongly against it. To the extent that you represent an otherwise non-vocal minority, I am intrigued by the strength of your conviction.

We live in a wealthy nation. The wealthiest the

world has ever known. In some ways, we define our culture (passively or actively) by what we do as a nation as expressed through congressional funding priorities. The National Endowment for the Arts is funded because it contributes a dimension to the quality of life we enjoy as Americans. Transportation is funded (even subsidized) because we value the economic vitality that it brings. The National Science Foundation is funded because it drives basic research that, history has shown, is a foundation of technological progress, especially where corporate R&D does not go. The Smithsonian is funded because we value the preservation of who and what we have been, to ourselves and to the world. The military is funded because we (as a nation) value above all else the real or perceived security it brings. The National Institutes of Health is funded because we care greatly about curing disease—and on and on. It is this mixture that defines America as a nation.

I suppose another way to set priorities would be to rank the problems of America (and of the world) and solve them in order, applying all resources at once. I believe this scenario is more resonant with your sense of what I should be doing with my life. But the history of the search for solutions does not support this contention. As I said, the most innovative solutions to problems commonly come from outside of the field—from people inspired by different priorities. The government knows this (primarily through fighting wars, and not through some deep insight into human nature) and values it through high investments in pure science, compared with the arts, for example.

No one has ever suggested that obtaining Hubble images is more important than feeding people. But this premise seems to feed your objections. The best of all worlds is to do it all. And, with all the flaws in the system, we nonetheless do it all better than anybody.

Again, thank you for your interest in my interview, and I do appreciate your comments in spite of (or perhaps because of) our points of disagreement.

Sincerely,
Neil deGrasse Tyson

Semper Fi

Thursday, March 14, 2019
Hi Neil,
*A lot has happened since I last wrote you.** *Not sure where to start. Some for the good but most for the bad.*

I guess with all things I'll start with the good. My career has been going amazing. I took a job flying "Flight For Life" and saved a few lives! I'm now back in Las Vegas flying as chief pilot for Maverick Helicopters doing tours and showing people how Earth was formed through some of the layers in the Grand Canyon. In that regards things have been great!

For the bad, I'm not sure where to start here. I've been through quite a lot. I know you don't know much about me from our few emails (other than I'm a big fan). I served six years in the Marine Corps and in that time I lost a few

* Jay's first of five emails to me was in 2013.

friends, one was my best friend. I felt like I paid a good price having been through that crazy time in my life. That's when I met my wife and we had our daughter. I was on Cloud 9! My wife worked for the Nevada test site as an engineer and we were perfect for each other.

She was diagnosed with breast cancer about four years ago. She battled it for three years like a champ but last year she lost her fight. I thought I would be prepared for it but broke me. If it weren't for our daughter Ella I'm not sure I would have recovered.

I just wanted to check in with ya and see how you were doing? Sorry if that email was a little drab. But I hope you're doing good! I'm still following you and I'll always have your back!

Your friend
Jay Scoble

Friday, March 15, 2019
Dear Jay,
When you assess all the parts and functions of the body, we should be amazed that human physiology works at all. So when parts fail, as they will for all of us at some point, or even when tragedy strikes, as in your lost friends in the Corps, not enough of us reflect on how amazing it was to have been alive in the first place.

Consider further that the Homo sapiens genome is capable of generating trillions of unique humans, which means most people who could ever exist will never even be born. So death is a kind of privilege of those few of us who have known life.

Such a cosmic perspective empowers me to celebrate every day I am alive. And I share it with you, as a kind of scientific solace on the life and death of your loved ones.

Peace to you,
Neil

Chapter 8

Tragedy

This chapter contains letters describing my firsthand account of the September 11, 2001, attacks on New York City's World Trade Center Twin Towers, written primarily to assuage the concerns of those who knew of my proximity to the danger. This chapter also includes a bit about conspiracy theories and a candid exchange with a mystic.

The Horror, the Horror*

Wednesday, September 12, 2001, 10:00am
Dear Family, Friends, & Colleagues,
My entire family is safe. We evacuated our lower-Manhattan residence at about noon yesterday and migrated north on foot to Grand Central Terminal (about three miles) where we took Metro North to

* A heavily forwarded email that became the subject of an article in the *Wall Street Journal* a week later on the widespread use of the World Wide Web to communicate the tragic news of the day.

the home of my parents in Westchester, from where I write this message.

We live four blocks from the World Trade Center, in view of both Towers, City Hall, and City Hall Park. I happened to be working at home yesterday. My wife went to work at 8:20am. I left at the same time to vote in NYC's Mayoral Primary. My 9-month old son was at home with our nanny. My 5-year old daughter was attending her second day of Kindergarten at PS-234, three blocks from World Trade Center. Lineup time in the yard was 8:40am in full view of WTC-1.

When the first plane hit at 8:50, they evacuated the school without incident. I noticed WTC-1 on fire in a high floor upon returning from voting, about 8:55am. Large crowds of onlookers were gathering along the base of City Hall Park as countless fire engines, police cars, and ambulances screamed past.

I went home, grabbed my camcorder, went out to the street and started filming. I consider myself to be emotionally strong. What I bore witness to, however, was especially upsetting, with indelible images of horror that will not soon leave my mind.

1. I first see WTC-1 on fire at a high floor. Not just flames coming out of some windows, but four or five entire floors on fire with smoke penetrating floors still higher.

Upsetting enough, but then . . .

2. Among the papers and melted steel fragments fluttering to the ground, I notice that some debris was falling distinctly differently. These weren't parts of the building that were falling. These were people, jumping from the windows, their bodies tumbling in rapid descent from the eightieth floor. I noticed about ten such falls, morbidly capturing three of them on tape.

Upsetting enough, but then . . .

3. A fiery explosion burst forth from a corner of WTC-2 about two thirds of the way up, perhaps the 60th floor. The fireball created an intense radiative impulse of heat from which we all had to turn our heads. From my vantage point, I could not see the plane that caused it, which hit 180 degrees on the other side of the building. Nor did I know at the time that a plane caused it. I first thought it was a bomb, but the explosion was not accompanied by the tell-tale acoustic shock-wave that rattles windows. This was simply a low frequency rumble.

   As it burst from the building's corner, the fireball was so large that it extended all the way across to WTC-1. The fact the building's corner exploded tells me that the ignited jet fuel got focused by the sides of the floor into which the second plane flew, meeting at the corner with increased explosive pressure. The flames were accompanied by countless thousands of sheets of paper that burst forth, fluttering to the ground as though every filing cabinet on multiple floors was emptied.

The fact that the second tower was now on fire made it clear to us all on the street that the first fire was no accident and that the WTC complex was under terrorist attack. Morbidly, I have the explosion on tape and the sounds from the horrified crowd surrounding me. At this point I stopped filming, and went back inside my apartment.

Upsetting enough, but then . . .

4. As more and more and more and more and more emergency vehicles descended on the World Trade Center, I hear a second explosion in WTC-2, then a loud, low-frequency rumble that precipitates the unthinkable—a collapse of all the floors above the point of explosion. First the top surface, containing the helipad, tips sideways in full view. Then the upper floors fall straight down in a demolition-style implosion, taking all lower floors with it, even those below the point of the explosion. A dense, thick dust cloud rises up in its place, which rapidly pours through the warren of streets that cross lower Manhattan.

I close all our windows and blinds. As the dust cloud engulfs my building, an eerie darkness surrounded us—the kind of darkness you experience before a severe thunderstorm. I look out the window and can see no more than about 12 inches away.

Upsetting enough, but then . . .

5. Outside my window, after about 15 minutes, visibility grows to about 100 yards, and I notice about an inch of white dust everywhere. That's when I realize that every single rescue vehicle that had parked itself at the base of the World Trade Center must now be buried under 110 collapsed floors of tangled debris, and multiple feet of dust. This collapse took out the entire first round of rescue efforts including what were surely hundreds of police officers, firefighters, and medics.

   As visibility increased and I could now see the blue sky, there was blue sky where WTC 2 used to be.

Upsetting enough, but then . . .

6. I decide it's time to get my daughter, who was taken by the parents of a friend of hers to a small office building, six blocks farther from the WTC than my apartment. As I dress for survival: boots, flashlight, wet towels, swim-

ming goggles, bicycle helmet, gloves, I hear another explosion followed by a now all-too-familiar rumble that signaled the collapse of WTC-1, the first of the two towers to have been hit. I saw the iconic antenna on this building descend straight down in an implosion twinning the first.

This dust cloud was darker, thicker and faster-moving than the first. When this round of dust reached my apartment, fifteen seconds after collapse, the sky turned dark as night, with visibility of no more than about a centimeter. It was getting harder to breathe in the apartment, but we were stable.

At this point I offer no hope of survival for any of the rescue personnel who were on the scene.

Upsetting enough, but then . . .

7. The cloud settles once again, now leaving a total of about three inches of dust outside my window. Another dark cloud of smoke now occupies the area where two 110-story buildings once stood. This cloud, however, was not the settling kind. It was smoke from ground-level fires. At this time the air in the apartment is getting harder and harder to breathe and it becomes clear that we should evacuate—especially with the likelihood of underground gas leaks. I load up my largest

backpack with survival items, put my son in our most nimble stroller and leave with our nanny, who then walks across the Brooklyn Bridge toward her home.

I go to where my daughter was held, which was upwind from all debris on a quiet street. She is in good spirits, but clearly upset. I have a crayon drawing of hers, sketched while waiting for me to arrive, which shows the Twin Towers with smoke and fire coming from them, as only a 5-year old could draw. "Daddy, why do you think the pilot drove his plane into the World Trade Center?" "Daddy, I wish this was all just a dream." "Daddy, if we can't return home tonight because of all the smoke, will my stuffed animals be okay?"

Upsetting enough, but then . . .

8. From the calm of an upholstered couch in the office where my daughter was kept, with my son under one arm and my daughter under the other, I realize that, fully loaded, each tower of the WTC holds 10,000 people. From what I witnessed, I have no reason to believe that any of them survived. In fact, I would not be surprised if the death toll reached 25–30,000. Beneath the Towers is an entire universe of six subterraneous levels containing scores of subway platforms, plus a hundred or so shops and restaurants. The Towers simply

collapsed into this hole—a hole large enough to have supplied the landfill for the World Financial Center across the West Side highway from the World Trade Center.

Upsetting enough, but then . . .

9. I realize that if the death toll is as high as I suspect, this incident is much, much worse than Pearl Harbor, where several thousand people died. It's more spectacularly tragic than the Titanic, the Hindenburg, Oklahoma City, car bombs, and airplane hijackings. The number of deaths in one four-hour period will be nearly half of the American death toll in all of Vietnam.*

I reconnected with my wife by 4 PM, meeting her just north of Union Square Park, before we hiked another mile north to Grand Central Terminal for our ride to Westchester, above New York City.

I will never be the same after yesterday, in ways that I cannot foresee. I suppose that my generation

* My worst fears for the death toll were much too high. I imagined at the time 25,000 deaths—the loss of two fully loaded, 107-story office buildings. But the Towers were far from full with people that early in the morning. The death toll from all three locations—NYC, DC, PA—came "only" to 2,998 people, with 2,606 of those dying in the World Trade Center itself.

now joins the ranks of those who lived through unspeakable horrors and survived to tell about it. How naive I was to believe that the world is fundamentally different from that of our ancestors, whose lives were changed by bearing witness to the 20th century's vilest acts of war.

Peace to you all,
Neil deGrasse Tyson
Hastings-on-Hudson, New York

Sunset on the World Trade Center

A love letter, appearing in the "City of Stars" special issue of Natural History *magazine.*

January 2002
Rising a quarter of a mile into the sky, the World Trade Center's twin towers were about five blocks tall.

I live four blocks from where they stood. I saw them ablaze. I saw them fall. All from my dining-room window, which, within ten seconds of each tower's collapse, offered less than one inch of visibility while the opaque dust cloud of pulverized concrete rolled by. From that same window, blue sky now appears where the twin towers used to be.

The World Trade Center was a veritable vertical universe. I think about it often. I think about the people who worked in the towers, the tourists who visited the observation deck, the diners at Windows on the World. I think of all those who lost their lives.

When I look hard for a peaceful way to remember the towers, I cannot help but think of them as observatories. On the top floor, you could type greetings into a computer that would transmit your message into space via the north tower's radio antenna, for all eavesdropping extraterrestrials to decode. The towers were so tall that for someone on the observation deck, the horizon was forty-five miles away. This distance was far enough along Earth's curved surface for the Sun to set two minutes later for the person on the observation deck than it did for someone on the ground floor. If you could have run up the stairs at one flight per second, you would literally have stopped the sunset. Alas, you'd eventually have run out of breath or run out of floors. In either case, at that moment you'd lose the Sun for the night, as it set gently below your horizon.

New York City's twin towers have lost the Sun forever. But I take comfort in knowing that the Sun will rise again each day, as it has done a trillion times before.

World Trade Center Anniversary

Wednesday, September 11, 2002
New York Times

To the Editor:
When I think of anniversaries, I think of occasions to remember people, places, and events that have been largely forgotten over the previous year. For me, however, not a day has gone by when I did not think of the World Trade Center and the thousands of lives lost in its rubble, just four blocks from my home. So maybe I'll use the anniversary as an excuse to try and think of something else for a day.

Neil deGrasse Tyson
New York City

Flags of Our Fathers

Friday, December 7, 2012
New York Times

To the Editor:
For most of my life I've wondered if the December 7, 1941, Japanese attack on Pearl Harbor, in which 2,400 Americans died, would ever fade emotionally—perhaps as the event became more of a distant memory, or as those who witnessed it passed away. On December 7, 1991, Pearl Harbor's fiftieth anniversary, I figured that

one remembers tragedy as long as a bigger, more recent tragedy does not arise to block your view through time. Indeed, ten years later, on December 7, 2001, a mere three months after three thousand Americans had died from the terrorist attacks of September 11, 2001, hardly any attention was given to Pearl Harbor, except perhaps as a yardstick for how to take measure of September 11th. If sustained peace and tranquility prevails, that will be good, but will carry the necessary consequence that my view of 9/11 itself will remain unobstructed, for better or for worse.

Neil deGrasse Tyson
New York City

Heavy Metal

Tuesday, March 31, 2009
Mr. Tyson,
It is a pleasure to be writing to you. I've become a big fan since I first saw you on The Daily Show *(maybe* The Colbert Report*). My girlfriend and I very much appreciate your views, humor, and approach to such interesting subject matter. The reason I write you today is regarding some of the disputed science of the 9/11 events. I know you were there on the day and if this is in anyway inappropriate or a subject you wish not to touch on, I apologize and respect that.*

My concern is the melting point of steel and if the 3 towers, including WTC-7 (the 3rd tower to collapse) could have been brought down the way they were without the use of controlled demolition. Richard Gage, the founder of Archi-

tects and Engineers For 9/11 Truth, has a very interesting traveling show which he uses to try and open up people to the idea that the Trade Center towers were indeed brought down by controlled demolition. I encourage you to see his show in person or speak with him if you ever have the chance.

Please let me know your opinion, whatever it may be, at your convenience. A subject like this could really use the opinion of respected mind like yourself. I hope all is well and thank you so much for your time!

Simon Naylor

Dear Mr. Naylor,
For any unique event, there will always be elements that one cannot explain, since they occur without precedent.

But one must always recognize the difference between knowing that something is true, knowing that something is not true, and not knowing one way or another. It's the not knowing part that leaves singular events susceptible to inventive accounts (especially from conspiracy theorists) of what may have happened.

And, of course, conspiracy theorists know the answers before they investigate, which contaminates their analysis—seducing them to accept what supports their thesis and reject or ignore or not notice that which conflicts with it. This psychological effect is well known in research communities, which is why peer-review is so important.

The controlled demolition hypothesis *required* that the towers collapse at a near gravitational free-fall.

The rapid fall of the towers was widely cited by the 9/11 deniers as evidence of this. I found the claim intriguing and tested it. From video coverage of the event, I timed the collapse of each tower. They in fact took about twice as long to fall when compared with free fall. This can be established from equations you learn in freshman physics.

I told this to a 9/11 denier on an impassioned email chain and he quickly wrote back, with dozens of people cc'ed, that I was lying and that I was in collaboration with the government.

Meanwhile, the 9/11 deniers are *not* citing the much-slower-than-free-fall of the towers as strong evidence against their case.

What tends to feed them are unexplained aspects of the day's events, stitching them together in ways that support their case. Forgetting of course that they are unexplained, and therefore support or deny nobody's case.

Sincerely,
Neil deGrasse Tyson

Symbolism, Myth, and Ritual

Sunday, November 15, 2009
Dear Neil deGrasse Tyson,
I hope you won't find the following questions too strange. What I am after is being able to answer, based on the research I've done into ancient and esoteric sources, whether there might be any merit to the idea that the attacks (as strange as this may sound) could have been coordinated with the move-

ments of certain celestial bodies in mind. To be able to write about this possibility critically I need an accurate sense of just where these bodies were on the day of and, more particularly, during the attacks, which started (for the purposes of my research) at 8:46am and concluded at 10:28am.

I have a strong interest in symbolism, myth and ritual, and thought I would try to bring a degree of academic seriousness and integrity to the conversation on the ritual aspect of human violence, however it might be seen to manifest. I would be interested in any thoughts you might have to share.

Thank you for any time you might be able to give this.

Sincerely,
Tom Breidenbach

Dear Tom,
People perennially over-interpret celestial events. The urge to link terrestrial affairs to cosmic phenomena is strong and deep.

Consider that an event can be rare but uninteresting. This happens all the time in the cosmos and fools people into assigning meaning when none is there to be had. For example, the exact crescent moon/Venus combination in the sky that appeared a couple of months ago will not repeat for 5,000 years. But there are 5,000 other juxtapositions of the crescent moon/Venus that will also not repeat for 5,000 years. Which means you get some sort of a crescent moon/Venus combination every year.

So when an event is declared rare, absent the context of the frequency of similar rare events, then superstitious people tend to assign irrational significance when none is there. The numerology that followed September 11, 2001 (regarding the date and year) looked quite fertile, until you realize that practically any date and year, in the hands of a determined person, will generate troves of numerical coincidences, giving the illusion that the date under study bears special significance.

Just words of caution as you look for otherwise unstated metaphysical significance of earthly events. Terrorist attacks often commemorate previous Earth-based events or attacks, without reference to the universe.

Neil

Chapter 9

To Believe or Not to Believe

The capacity of the human mind to believe, in the absence of tangible evidence, knows no bounds. Those who wrote to me about their beliefs, in almost all cases, were trying to win me over to their side, but were also authentically curious. As an educator, I had no hesitation engaging them, but I'm also genuinely curious about all the ways a person's mind can be wired for thought in our perennial attempt to make sense of the world.

The Eye of God

Friday, May 20, 2005
From the Internet . . .

God is looking back in the other end of the telescope?
NASA CALLS IT THE EYE OF GOD
This was entirely too cool not to share!
This is a real picture . . .

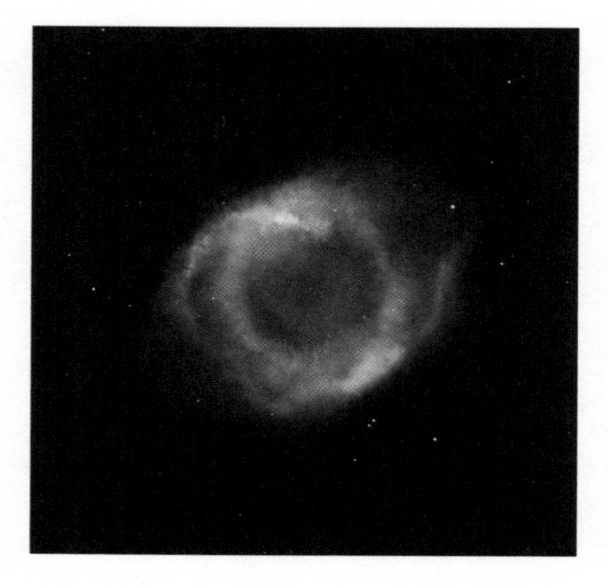

Could that be an actual photo?

Fond regards to you & your family,
*Niki Branford**

Hi Niki,
Real photo. Real object in our Milky Way galaxy called the "Helix Nebula," a.k.a. NGC7293. Taken by the Hubble Space Telescope.

The urge to look up, see something beautiful, and call it God is strong. In the first century AD, the famous astronomer and mathematician Claudius Ptolemy felt that way as he studied the motions of the planets against the background stars and penned:

* Correspondence with the sister of a lifelong friend.

> "When I trace at my pleasure the windings to and fro of the Heavenly Bodies, I no longer touch earth with my feet. I stand in the presence of Zeus himself, and take my fill of ambrosia."*

One of my favorite quotes of all time.

What has always intrigued me, however, are all the other things that go on in nature, in this same universe, yet people do not feel compelled to wax poetic about the majesty of God. Like rapid-growth cancer cells, fatal birth defects, killer tsunamis, killer earthquakes, killer volcanoes, killer hurricanes, killer asteroids, the Ebola virus, lethal parasites, malaria-carrying mosquitoes, plague-carrying rats, Lyme disease, heart disease, stroke, appendicitis, species extinction . . . the list is long—practically endless. And how about the equally long list of ghastly things to look at in nature? Such as the close-up image of the dust-mite, or a close-up view of the underbelly of a tarantula, or the sucking jaws of a leech, or the slimy trail of a banana slug, or the flea-infested underbelly of a dog . . . and so forth.

So when I see the Helix nebula, I simply see a strikingly beautiful part of our galaxy, but with no particular urge to credit or blame anybody for it.

Neil

* Owen Gingerich, *The Eye of Heaven: Ptolemy, Copernicus, Kepler* (Washington, DC: American Institute of Physics, 1993), 55. Epigram handwritten by Ptolemy in the manuscript of his *Almagest* (ca. AD 150).

Thinking for Yourself

In December 2011, on the internet chat universe of Reddit, I was asked what books should be read by every intelligent person on the planet. I replied with a ranked list of eight volumes, each accompanied by a short phrase explaining why. I ranked the Bible No. 1, but my comment irked many a believer. A few years later, catching up with the comments in the thread, I posted a response.

The list . . .

1. *The Bible*
   " . . . to learn that it's easier to be told by others what to think and believe than it is to think for yourself"

2. *The System of the World* by Isaac Newton
   " . . . to learn that the universe is a knowable place."

3. *On the Origin of Species* by Charles Darwin
   " . . . to learn of our kinship with all other life on Earth."

4. *Gulliver's Travels* by Jonathan Swift
   " . . . to learn, among other satirical lessons, that most of the time humans are Yahoos."

5. *The Age of Reason* by Thomas Paine
"... to learn how the power of rational thought is the primary source of freedom in the world."

6. *The Wealth of Nations* by Adam Smith
"... to learn that capitalism is an economy of greed, a force of nature unto itself."

7. *The Art of War* by Sun Tzu
"... to learn that the act of killing fellow humans can be raised to an art."

8. *The Prince* by Machiavelli
"... to learn that people not in power will do all they can to acquire it, and people in power will do all they can to keep it."

Why is my comment on the Bible so non-complimentary?

1. The Judeo-Christian Bible is likely the greatest (single) source of tribalistic conflict the world has ever known. I have no problems with those who assert that people's warped interpretations are what is reprehensible and not the Bible itself. But that does not absolve people who behave in these ways not from free thought but from Biblical passages asserted to be divine in origin. This conduct germinates hierarchies based on unassailable authority—

dogma. When you are under the influence of dogma, you say, do, and think what others tell you to. And that's always easier than thinking for yourself or resisting the powers that established the dogma in the first place.

2. Of course religion is not the sole source of dogma in the world. There's political dogma, as well as cultural & ethnic dogma. There's even, on occasion, scientific dogma. But science contains the methods and tools within itself to ferret it out, so dogma in science doesn't last long when it arises. Consider also that scientists hardly ever wield power. So when science becomes dogma in a country, it's usually because a political system that is itself dogma has adopted it. Nazi Germany and communist Lysenko Russia are, perhaps, the best example of this.

3. A reminder that the task I was given was to list books I felt that an educated person should read—books that would impart insight into the human condition and the trajectories of civilization that derive from it. The conduct of people who have tribalized after reading the Bible—people who have participated in a kind of "group-think"—are responsible for shaping large swaths of Western human history. All of which leads to my single sentence that "it is easier to be told by others what to think and believe than it is to think for yourself."

For these reasons, I maintain the intent and significance of that sentence.

Respectfully Submitted
Neil deGrasse Tyson, New York City

God and the Afterlife

Wednesday, November 29, 2006
Hello Dr. Tyson,
My question to you is (and I'm sure it will seem like a loaded one), do you believe in a supernatural being such as God and the prospects of an afterlife? If not, then what, or how, do you explain it (the concept of religion and why some believe) to your children?

As I have pondered this question for some time, I asked myself that if God and an afterlife is not real, then why has this concept become so fundamental to human societies from the beginning of their existence.

Your time and answers will be appreciated, but in the end, I will probably still say my prayers because the little investment of faith can't hurt—just in case He's up there and there is something more after my body has passed away.

Webster Baker

Dear Mr. Baker,
I've yet to be convinced by anything I see on Earth or in the universe that anyone or any intelligent entity is in charge.

I teach my children about all the world's major religions. Not pejoratively but anthropologically, which is, I think, a sensible way to broach comparative religion. In this way, they know that while there are multiple belief systems in the world regarding God and gods, there is only one science, and that science is the same no matter your birthplace, on Earth or anywhere else in the cosmos.

I do not know whether or not God is real. I simply know that the people who cite evidence in favor of God have overlooked the preponderance of evidence against it.

Other widespread and timeless activities of human societies include war and infidelity and power struggles and slavery and exploitation. Just because something endures within and across cultures does not mean it is good or correct or the right thing to do for the future.

As for the urge to believe in an afterlife, note that for most of the history of life on Earth you did not exist. A condition that continued right up until your birth. That is not a hard thought to consider. Nor is it depressing. You simply had no existence or awareness of anything at all. It should therefore not be hard to consider the likelihood that the state of death is no different.

As for saying your prayers, just in case, it reminds me of a story about the horseshoe that hung in Niels Bohr's office. This famous physicist was asked why he, a man of science, believed in such superstitious things. He's rumored to have replied, "They tell me it works even if you *don't* believe in it."

Sincerely,
Neil deGrasse Tyson

Seeing Eye to Eye

Thursday, September 30, 2004

Dear Neil,

Hello. My name is Tom. I saw you on the PBS show Origins, *where you discussed the beginning of the universe. From as far back as I can remember, I have been fascinated by space, stars, and the moon. I am an amateur radio operator, and currently work for a company that specializes in HAM radio amplifiers and equipment.*

I do have to disagree with the whole theory of evolution of this universe and here is why:

I am a Christian who believes that God created this universe and actually spoke it into being. I can believe that perhaps life could possibly exist elsewhere. I am open to this. The Bible does not mention it, but then again, it does not mention dinosaurs either. You see, Earth was very different in the time of Adam and Eve, before sin entered. There was no sickness or death. Animals did not attack or eat other animals. There were no hurricanes, tornadoes, earthquakes, and other things.

I know you think this is crazy. I was told by my professor that science and God could not mix, but science cannot happen without God.

I hope that you and I can possibly see eye to eye with our different viewpoints of origin, for we both love science and nature.

Sincerely,
Tom Rodenstock★

★ Name changed by request.

Dear Tom,
Thank you for your comments. The subject of origins never fails to trigger all manner of reactions. In the end, people tend to place their own filter on it, in ways that best resonate with their personal worldview.

Your point of view derives, of course, from the Judeo-Christian Bible (the Old Testament). The problem here is that plenty of people in the world who are religious believe something else, expressing no less confidence than you do in their specific belief system. The Animists, Buddhists, Confucians, Hindus, Jews, Muslims, Shintoists, Voduns, etc. are each as certain as you that their beliefs are the moral, correct, and only ones to hold. Not to mention the countless sects within Christianity itself, whose beliefs and traditions all differ in important ways: Anglicans, Baptists, Catholics, Episcopalians, Jehovah's Witnesses, Lutherans, Mormons, Presbyterians, Seventh Day Adventists, and so forth. In the past (and even in the present) differences can drive adherents to commit murder against another sect in the name of their faith.

Science, on the other hand, is a system of knowing and discovery that is decoupled from your nationality or birthplace or ancestry or politics or who you worship. It's a system of knowing the natural world that is immune to opinion, but not to experiment.

When you invoke the methods and tools of science, you come up with the origins story as conveyed by scientists. You don't come up with any faith-based description of the natural world. Had that been the case, then scientists would have been mining religious writings since antiquity for insights into how the physical world works.

Again, thank you for your interest, and continued best wishes for your HAMming it up on the airwaves.

Sincerely,
Neil deGrasse Tyson

The Bible Tells Me So

Brandon Fibbs, a former devout Christian turned atheist, engaged a former professor from when he attended bible college. Certain the Bible is correct on all matters, and that anything disagreeing with the Bible amounts to a liberal conspiracy, the professor actively denies global warming, evolution, the Big Bang, and other findings from the frontier of science. Fibbs, a formidable writer and commentator, shared his 1,500-word reply with me to solicit my reaction. What follows is my response to his full-scale attack on his former professor.

Sunday, February 14, 2010
Brandon,
Your polemic is tight, unforgiving, and well-informed. Would probably be a waste on someone other than a former professor. My bet is that you are older now than he was when he was your professor. Is that correct?

Personally, I try to spend twice as much time to make things half as long. You don't want to be guilty of a Shakespearean, "Methinks the Gentlemen doth protest too much." Another point I take to heart is the adage, "When an argument last longer than five minutes then both sides are wrong."

About global warming and heavy snowfalls, I continue to find it odd that people equate snow accumulations with cold. The heaviest snowfalls tend to occur between 22 and 32 degrees Fahrenheit. At these "warmer" temperatures, water crystals grow larger, are stickier, and accumulate much faster on the ground. So huge snowfalls are indications of warm snowstorms not cold ones.

Also, try to avoid the word proof. It's commonly ascribed to what scientists do. But it somewhat misrepresents what goes on in the process of discovery and confirmation. And it leaves you susceptible to people saying that "Scientists once *proved* that A is true, but now they say B is true." This highlights the modern distinction between the words hypothesis and theory.

Scientists never "prove" anything. This word has a specific application in mathematics, but in science what we do is demonstrate, with sufficient experiments, that a consensus exists and further evidence in support of an idea would be a waste of effort or funding since other pressing questions remain unanswered. When such an experimental consensus emerges *the results will never one day be shown to be wrong.* In the modern era of science (the past 400 years) all that happens is that a bigger truth emerges that enclosed the previous ideas and experiments in a deeper understanding.

For ideas in progress, we are now using the word "hypothesis" rather than Theory, reserving the word theory for big ideas that provide broad and deep understandings of the operations of nature. Quantum Theory, Relativity Theory, Evolution Theory. Some theories of the 19th century still retain the word "Law," back when that word was common: Laws of Gravity, Laws of Thermodynamics, etc. Today they would be called theories.

Neil

p.s. Your writing is potent, but you never want to win an argument for being a better writer, or because you have a bigger vocabulary than your opponent. In this way, the strength of the argument rests on the strength of the argument rather than the strength of your literacy.

A Piece of Pi

Sunday, November 28, 2004
Dear Neil,
You mention pi in a recent article of yours. For many years nearly all the books on the history of mathematics stated that in the Old Testament of the Bible, the value of pi is given as 3, a rather poor approximation. Yet recent "detective work" shows otherwise.

One always relishes the notion that a hidden code can reveal long lost secrets. There are two places in the Bible where the same sentence appears, identical in every way except for one word, which is spelled differently in the two citations.

Reverting to the original Hebrew, in I Kings 7:23 it was written as הוק*, whereas in 2 Chronicles 4:2 it was written* *as וק. Elijah applied the ancient biblical analysis technique (still used by Talmudic scholars today) called gematria, where the Hebrew letters are given their appropriate numerical values according to their sequence in the Hebrew alphabet, to the two spellings of the word for "line measure" and found the following: the letter values are: ק = 100, ו = 6 and ה = 5. Therefore, the spelling for "line measure" in 1 Kings 7:23 is הוק = 5 + 6 + 100 = 111, while in 2 Chronicles 4:2 the spelling וק = 6 + 100 = 106. Using gematria in an accepted way, he then took the ratio of these two values: (to four decimal places), which he considered the necessary "correction factor." By multiplying the Bible's apparent value (3) of pi by this factor, one gets 3.1416, which is pi correct to four decimal places!*

"Wow!" is a common reaction. Such accuracy is quite astonishing for ancient times. Moreover, remember how just getting pi = 3.14 using string measurements was quite a feat. Now imagine getting pi accurate to four decimal places. We would contend that this would be nearly impossible with typical string measurements. Try it if you need convincing.

Dr. Alfred S. Posamentier, Dean, School of Education
The City College of New York

Dear Alfred,

No need for you to be rapt by Talmudic Kabbalist numerology. To know, in advance, the answer you are looking for, and then to manipulate a set of preexisting numbers in search of the connection, is an old, attractive, but discredited means of knowing the world. The real measure of numerology's value (if it had any) would be to do all this in advance, and then *predict* the value for pi (or anything else). But this has never happened using numerology. That's because there is an almost-infinite number of ways you can combine numbers to get other numbers. And if you do not know in advance what you are after, you will most surely be doing calculations that have nothing to do with anything.

The "power" of numerology is seductive indeed. As just one example of many, the attacks on September 11, 2001, spawned endless excursions in numerology, all pointing to some deep meaning for the time, date, number of hijackers, number of letters in their names, and so forth.

Problem is, this information did not exist *before* the attacks, thereby missing an opportunity to predict them. That's because after-the-fact numerological deductions can be made for any event of any date of any year (you would just combine the numbers in another, equally sensible way), yet the results appear to the eye and the mind as a magical, even mystical connection.

Other hot-beds for numerology include the Kennedy assassination, the shape and proportions of the Egyptian pyramids, the end of the world, the attack on Pearl Harbor, and D-Day.

So have fun, but do so knowing that numerology is granting access to entertainment rather than to access reality.

Neil

Buddhist

Friday, August 28, 2009
Hello Dr. Tyson,
I want to say that I enjoy your videos very much. And yes I consider myself a religious person. How come you guys don't pick on the Buddhists? It's only the Christians, Jews and Muslims. If you haven't guessed I am of Buddhist beliefs. Not important, but it makes the joke more funny.

I want my children exposed to other ideas and let them make a decision for themselves as to what they want to believe. I only want to teach them compassion, and if they follow that through science or religion, good for them.

Keep up your great work sir.
Todd Baxter

Dear Mr. Baxter,
In my writings (from which the videos are derived) I reference only those who want to bring religious philosophy into the science classroom. This behavior is common for Protestant fundamentalists and largely unheard of in America among Buddhists, Jews, or Muslims, for that matter.

Note also that not all belief systems are equal. Most are demonstrably false. The notion that all belief systems are equal is evidence for rampant scientific illiteracy in the nation.

And you care about compassion—as we all should. But to be devout in one religion almost always requires that one reject all other religions. Compassion is the last thing you find during Holy Wars. And, of course, the most prominent stories of the Old Testament show anything but compassion.

Thank you,
Neil deGrasse Tyson

Open Mind

Thursday, August 13, 2009
Dear Dr. Tyson,
I respect you very much. I also love my church. I am very confused. I want to ask one question—as a man of science, you are supposed to have an open mind—is there ANY chance at all that the Earth is only five or six thousand years old?

I would just say that if there is no God I would/will feel awfully alone and insignificant.

Kevin Carrol

Dear Kevin,
Zero chance that Earth is five or six thousand years old.

As I say often, if you use religious texts to predict future knowledge of the physical universe, you will get the wrong answer. But it's not for want of trying. More accurately stated: every previous attempt at this exercise has failed.

Consider instead Galileo's dictum*:

* Stillman Drake, *Discoveries and Opinions of Galileo* (New York: Anchor, 1957), 173.

> "In my mind God wrote two books. The first book is the Bible, where humans can find the answers to their questions on values and morals. The second book of God is the book of nature, which allows humans to use observation and experiment to answer our own questions about the universe."

Galileo was a religious man, but was nonetheless compelled to say:

> "I do not feel obliged to believe that the same God who has endowed us with senses, reason and intellect has intended us to forego their use and by some other means to give us knowledge which we can attain by them. He would not require us to deny sense and reason in physical matters which are set before our eyes and minds by direct experience or necessary demonstrations."

Just to be clear, the existence or non-existence of God has nothing to do with the age of the Earth. Most (> 80%, is my guess) of western religious people recognize this. Those who link Earth's age with whether or not God exists are a small minority of the religious community. They just happen to be louder than most others, giving the (false) impression that they are mainstream and represent the majority. There are extensive religious organizations that have published statements in support of Evolution, which requires accepting a very old Earth.

Good luck in your explorations.

Sincerely,
Neil deGrasse Tyson

Proof

Monday, September 19, 2005 thru
Monday, May 8, 2006
Hello,
I know you are a busy man and I hope you would reply to my humble email. Seeing you on TV presents me with mixed emotions.

First, I'm happy to see a fellow black man talking about science on such a wonderful and popular TV program such as PBS Nova *(one of my favorite programs). We certainly need to see more blacks in the scientific field, and* Nova *is a great place for it. I'm an electronics engineer myself (started two engineering companies) and I love science very much.*

On the other hand, I feel sorry that you don't seem to believe in God, and think that everything came into existence by random chance. As an electronics engineer, I design complex products all the time for a living, and I know how hard it is to design something and make it work. Every last detail must be thought out completely ahead of time or else all you get is fire and smoke, or get sued. I can see no way that something as complex as a human being, DNA, the universe with all its robustness could have made itself from random chance or from a Big Bang.

I was wondering why you don't believe that the universe was created by God? How could anybody believe that such a complex universe could be created by random chance? From personal research, I have found so many reasons why, scientifically, God must exist.

A lot of scientists panic that if people start believing that God designed everything, then we would stop trying to discover things because we already know who designed it.

But knowing the designer is God would make you want to discover stuff even more. I know a lot of people say that things are badly designed, but I don't think so. In a physical world there are limits to everything and compromises, no matter what you do (even God), physical things cannot be perfectly designed.

For example, how can anything physical be really perfect? Can it survive every attack, can I throw it into the Sun and it still comes out unscratched? Can I keep it in water for a year and it still can live? Can I keep it in a volcano and it still lives? Can I pour some toxic waste on it and it doesn't die? Can I hit it with some bird flu combination, AIDS and cancer and it just bounds off? Nothing can withstand all that; even if it could, there is always something else in the universe that can kill it.

I believe God already knows this, so no matter what he did, there would be something in the universe that could kill his physical creation, so why bother making it bullet proof. Anything physical will be mortal no matter what. This is what it means to be physical. Comes with the territory.

Thanks for listening to my ramblings!
Nigel Smith

Dear Nigel,

The modern intelligent design movement, as expressed in the Discovery Institute webpages and by leading ID proponents in the 2005 court case in Dover, Pennsylvania, invokes intelligent design only when the thing being described is unknown (e.g. the origin of life).

If you personally declare that things we do understand (and typically can control or influence) are also the work of an intelligent designer, then you operate with no restrictions on what to discover next.

About good versus bad design—to assert that one must survive a direct hit by a million-ton meteor goes far beyond any example that I give. Anybody with that ability would be "over designed" because such a danger is extremely rare. But choking to death is common in our species. So is drowning. So is childhood leukemia. So are (most) birth defects, etc. And no engineer, in his or her right mind, would ever design a system that ingested liquids, solids, communicated, and respirated through the same orifice. So you ask where to draw the line? Any rational person would put meteor strikes far to one side of the line, and choking to death far on the other side.

I am not in denial of good design. Good design is obvious when it's there—the opposable thumb. Stereo-vision. Speech. Ball sockets (shoulders and hips). The shape and strength of our skull, to name a few. But you are in denial of bad design, not because it is not there, but because it falls outside of your religious philosophy, and are thus blind to it. By the way, you are not alone. This has been going on for centuries. And there is an entire field of religious philosophy called "apologetics" that carries on with this behavior. Its proponents are called "apologists."

What they do is construct arguments to counter criticisms of Bible passages by loosely interpreting the literal word, thereby leaving room for the Bible to not sound as though it contradicts empirical findings of the natural world. A clean example of this is the fact that nowhere in the Bible is Earth referenced as a 3D object. And everywhere a reference is made, Earth is simply flat—typically drawn as a circle, with Jerusalem in the center, and the land masses surrounded by water in every direction on the horizon, as is clearly stated in many places and consistent with understandings of the known world of the day. The apologists then cite the Bible passage containing the words "circle of the Earth," and assert that Circle means Sphere. But, in fact, people of the day knew full well the difference between the two.

So therein lies a non-convergent point of our conversation: You already know where you want to land. And God is there to design it. I have no idea where I am going to land. And if there is a God of unwavering intelligence, this fact is simply not evident in the book of nature to the unbiased observer.

Natural selection never claims perfect design, or even good design, only a design that is more effective than that of a competing species, allowing survival long enough to reproduce. Nothing else matters to the process.

Further, I never said the universe was not designed.

I simply said that if it was designed, then there is ample (and widely ignored) evidence to demonstrate the designer's blunders right alongside all those things that are marvelous.

Sincerely,
Neil deGrasse Tyson

Meaning in Life

In December 2007, Mark, an inmate at the Kentucky State Reformatory, asked perhaps the deepest religious question of them all—If there is no God, then how can life have meaning? Who should care if you ever lived? Why would it matter if any one of us were Stalin or Einstein? Hitler or Mother Teresa? I offered Mark an answer, but with no premise that it's the only answer.

Dear Mark,
Often people, especially religious people, look outside themselves to find the meaning of life—in scripture, in messages from religious leaders, religious relics, and so forth. When you do this, and do it for your entire life, it becomes hard to imagine life without this kind of spiritual structure constructed for you and around you.

But suppose you instead looked within yourself? By doing so, it's not hard to find meaning in life by doing meaningful things—caring for others less fortunate than you, raising children, accomplishing hard tasks that give you physical, intellectual, or emotional fulfillment. The drive to do this, without any reference to religious texts at all, can be quite fulfilling. My personal goal is to leave the world a little better off for my having lived in it. The prospect of making this come true drives my work habits daily.

For some people, the search for meaning leads to violence, abuse of others, and crime. These people range from selfish to misanthropic. But those traits are not the unique province of non-religious people. The world is no stranger to religious warfare—with abject slaughter of countless innocents in the name of one god or another. So your supposition that one needs God to behave or to give meaning to life—while it may be true for many people—is certainly not a pre-requisite to a fulfilling, law-abiding life.

I add that if I help an old lady across the street, it's because she needs help and I can provide it, not because I am expecting rewards for having done so, either in this life or in heaven. My motivation is simply to create meaning and self worth—not only in my own life, but in the lives of others.

Finally, deeply religious people sometimes ask, "Without God, why should people be civil to one another?" or "What's to stop people from committing crimes or even murder in the absence of divine judgment?" There is a simple answer to this: prison. That's why laws exist, to restrain offensive behavior from person to person and between person and property. This recipe works for most people. In fact, in Europe, there are whole countries (e.g. Switzerland, The Netherlands, England, France, Sweden) where religion plays hardly any role in politics, culture, business, or family, yet they enjoy *far lower* rates of violent crime than in America, where more than 9 in 10 people say they are religious. In these other countries, the fraction of people who say they are religious is typically less than 1 in 10.

So rest well, whether or not you are religious, reflecting on the fact that in most western societies, religion is an aspect of culture, and not culture itself.

Best to you,
Neil deGrasse Tyson

IV.

Kairos

A propitious moment
for decision or action.

Chapter 10

School Days

A time and a place to learn new things, and to establish the intellectual taproots of life.

A Teacher, a Student, and a Church-State Dispute

A public middle school student recorded his science teacher giving Bible accounts of the natural world. He went public and it made headlines. Normally silent on such matters, I had to jump in, with a letter to the editor of the New York Times.

Thursday, December 21, 2006
New York Times

To the Editor:
People cited violation of the First Amendment when a New Jersey schoolteacher asserted that evolution and the Big Bang are not scientific and that Noah's ark carried dinosaurs.

This case is not about the need to separate church and state; it's about the need to separate ignorant, scientifically illiterate people from the ranks of teachers.

Neil deGrasse Tyson
New York City

Space Cadet

In April 2008, Ronald Ward, an African-American middle schooler with intense interest in space since age six, sought my guidance for his upcoming science fair project. He'd been to Space Camp multiple times. He wanted to be a pilot or an astronaut and every Sunday launches his own model rockets with his dad. He also happens to suffer from a seizure disorder, which made him the brunt of teasing by his fellow students, and may force him to alter his dreams of being an aviator. They called him names like "space cadet," "nerd," and "geek" and told him he will never become a scientist, mathematician, or engineer, which hurt his feelings.*

If he created a winning science fair project, he wondered whether the kids in school would start being nice to him. He further wondered whether kids made fun of me when I was in middle school.

Dear Ronald,

Thank you for that enthusiastic, personal note.

In my circles, we assign the playful title of "Space Cadet" with pride. And to be called a "Geek" is practically a badge of honor. Remember that one of the

* Name changed.

richest people in the world, Bill Gates, is a card-carrying Geek. So too is Mike Griffin, the head of NASA. As am I. So when your classmates make fun of you for being enthusiastic about everything aerospace, just remember that there are hundreds of thousands of us out there who understand you. And never forget that the only people who are any good at what they do in life are those who are passionate and persistent.

About your occasional seizures—they will certainly prevent you from becoming an astronaut—as do many common medical conditions, including most chronic ailments that require prescription medicine to overcome. But they will not prevent you from being smart, from being a mathematician, from being an engineer, from being a scientist, or from being the designer of airplanes and spacecraft used on the frontier of discovery.

Remember that for every astronaut in space, there are a thousand scientists and engineers who got 'em there.

I noticed from your return address that you live in the Rockies. Colorado Springs happens to be the headquarters of the Space Foundation—a kind of center of the universe for everything space. Among the many things they do is track how space technology becomes everyday products. I strongly recommend you visit them. If you do, I bet they will send you home with a box of cool stuff—pens, posters, pins, paperweights, and other ponderables that you can use in your science fair project. I know, because I served on their Board of Directors, and every time

I visited their headquarters, I went home with a box full of cool stuff myself.

If you visit the Space Foundation, you will surely get to hang out, even if only briefly, with others who had flatly ignored the bone-headed peer pressure of their classmates.

Best to you, on Earth and in the universe.
Neil deGrasse Tyson

Elementary Curiosity

Friday, April 10, 2009
Dear Neil deGrasse Tyson,
I think it's cool that you write a lot of books about the universe. I want to read them someday. I want to be an astrophysicist when I grow up, too. I am in first grade and am doing a project on my living hero. Can you answer these questions for me?

Thank you,
Gabe Mopps

1. *Do you know what causes the gravitational pull of planets and their moons?*

Hi Gabe,
Gravity remains a mysterious force in the universe. When objects wander near another object's field of gravity we invoke Einstein's General Theory of Relativity, which says that gravity curves space and time.

Objects simply follow those curves as they move. But aside from that, nobody knows what gravity actually is.

2. *Is it really hard to research about black holes because they are invisible?*

Yes. So we study what effect black holes have on their surrounding regions. Black holes do things to space and matter and energy that no other objects do. That's how we can find these invisible monsters in the universe. Kind of like seeing a bear's foot print in the snow, which tells you a bear was there even if you did not see the bear itself.

3. *How do you do your research for your books on these ideas?*

Read, Read, Read. Think. Think. Think. Read, Read, Read.

4. *I find all of these things really interesting.*

So do I.

5. *I heard that you might be head of NASA.*

Heard the same thing. Just rumors.

Your friend,
Gabe Mopps

Thanks, Gabe, for your interest.

And as we say in the universe, keep looking up!

Neil

Look But Don't Touch

Tuesday, February 5, 2008

Mr. Tyson,

I am 13 and I want to be an environmental engineer. But with space being the last frontier, it is good to be educated about space and nature too.

I do have one question:

Isn't it horrible not being able to touch the stuff you look at? All you can really do is use your eyes from light years away. It must be frustrating not being able to be close enough to use your hands.

Sincerely,

Marc Jaruzel

Dear Marc,

Yes, it can be frustrating to not get one's hands on the subject of interest. But we learn in astrophysics that the telescope is not only as good as the hands, in many ways, it's better.

Besides, who wants to touch a quasar? Or a black hole? That would not be a particularly safe thing to do.

Sincerely,

Neil

To Know

Tuesday, April 7, 2009
How do you know what you know?

David Lunianski

Dear David,
I was in school until I was 32. And since then I read a lot. School is not only a place to learn, but a place to learn how to learn. And at its best, school should be a place to stimulate a lifetime of curiosity.

In addition, as often as I can, I seek out people smarter than I am to talk to and hang out with. My wife, for example, has a PhD in mathematical physics. She knows a ton more stuff than I do about all kinds of things. And I won't have it any other way.

Neil deGrasse Tyson

Stigma

Thursday, July 24, 2008
Dear Dr. Tyson
I read with interest your observation (in the July 7, 2008 issue of Time *magazine) that improving student performance in science and math requires removing the stigma associated with the study of these subjects.*

After many years of observation, I strongly believe that a major cause of this poor performance is the low level of respect the media and society show to those who excel in science and math. After all, why would a student strive to excel in a subject that possesses little perceived value? For example, a quick reading of recent newspaper articles finds references to various people by titles such as "chef," "officer," "doctor," "ranger," etc. In fact, in the same July 7 issue of Time *magazine in which your observations appeared, no "Dr." appeared before the name of Neil deGrasse Tyson.*

As a scientist with a Ph.D. in theoretical physics who has taught several thousand students over a 35-year period at the University of Minnesota, I have had many conversations with students about this subject. And, during these conversations, a scientist's relatively low position within society was frequently cited as a reason (in addition to subject-matter difficulty) to avoid the study of science and pursue other professions that possess a higher "societal approval factor."

As the most visible member of the scientific community, you are in a superb position to begin the process of altering the manner in which society views scientists.

Thanks for your time.
Dr. Robert Cassola

Dear Dr. Cassola,
Thanks for sharing your note about the existence or absence of respect that scientists get from the public. You make an intriguing point, but certain (repeatable) surveys, as well as some anecdotal cases I can cite, do not agree with your contention, or rather, would argue that whatever needs to be fixed, is not traceable to titles in any important way.

From Salary.com, we get a glimpse at today's top

ten most respected professions. Of course, forty years ago, soldier and policeman would have been nowhere on that list, so times have changed for them too. And as expected, lawyers, politicians, and salesmen are absent.

1. Doctor
2. Soldier
3. Teacher
4. Firefighter
5. CEO
6. Scientist
7. Engineer
8. Police Officer
9. Architect
10. Accountant

While other surveys vary slightly, scientist, as a profession, has been in the top ten for at least 30 years.

Over the decades, there has been a real phase shift in how scientists are portrayed in movies and television. The mad scientist is a fading icon. In television, the shows *CSI: Crime Scene Investigation* and *NUMB3RS* are hit prime-time network crime dramas that feature socialized, attractive, brilliant scientists (chemists, mathematicians, physicists, biologists) in leading roles. In fact, the enrollment of girls in college chemistry and mathematics has increased dramatically over the years these shows have been hits—e.g. today, 48% of all mathematics majors in college are women.

Recent American Institute of Physics (AIP) data

reveal the median 12-month salary of senior professional scientists (in research academia or industry) to be twice that of the median national household income.

In my life experience, shedding the "Dr." title helps to dissolve a communication barrier and makes people want to learn more from you—provided that your pedagogical message is one that empowers the listener's capacity to think. If you accomplish this, they will beat a path to your door regardless of the title.

And as you know, unlike in the social sciences, in your and my professions, titles are omitted from published research papers—a tradition I have always enjoyed. I see it as a tacit recognition that, for example, an untitled graduate student might have an idea as important as that of a more senior researcher, and a reader of the paper would not necessarily know who was whom.

That being said, about 60% of my media interviews (print and broadcast) do invoke the title "Dr." and are quite respectful about it when they do. But in both cases—when they do and when they don't—they keep coming back to learn more about science, which is, for me, the best measure of respect there is.

There is more quality science documentary programming on television today than ever before. When you combine the broadcast calendars of PBS, Discovery Networks (including their Science Channel), National Geographic, the History Channel, and intermittent network specials on selected science subjects, the public's exposure, appreciation, and appetite for science has grown exponentially over the years.

Of course, none of this addresses the perennial and paradoxical fact of our low test scores and other metrics of poor performance relative to that of other developed nations. But it would be hard to pin that blame on titles not being used for scientists.

So, while your concerns are sensible and precise, the above information does not support them, and in particular, argues for an opposite trend. A good thing.

Thanks for your interest.

Neil

Not a Shadow of Doubt

Tuesday, June 30, 2009

Dear Dr. Tyson,

I am an Indiana State Police Officer, a big fan of scientific endeavors, and most of all, a fan of yours. I know you are a mega scientist and celebrity, but I wondered if you would speak to me about how I could use scientific applications (ie. observation techniques, crash damage investigation, investigative techniques/discoveries) in the field (on the road—not forensics). I like your thought process and would like to know if you would share a "layman" approach into how you perceive the world. My goal is to become a better Police Officer/Investigator, but at times use a different approach to get the same ends.

Maybe one day, when you are in the Chicagoland area, I can meet you personally.

Lawrence McFarrin

Dear Officer McFarrin,
Thanks for your note about trying to use science in your work. Of course the hit TV show, *CSI*, in its multiple incarnations (*CSI: New York*; *CSI: Miami*; *CSI: Cyber*), is all about using science to solve crimes, although they usually have one or two dead bodies to deal with along the way, and everyone not dead is otherwise good-looking.

In your specific case, I offer a possibly unorthodox reflection . . .

What you should do is not learn about how to apply the laws of physics to police work. Instead, you should learn the laws of physics—"Physics 101" at a local community college or university. Community colleges, as you surely know, tend to offer course schedules that knowingly accommodate working people, so they may be your best option.

When you learn about motion and gravity and forces and acceleration and statics and thermodynamics and light and electricity, then the ways and means of how they apply to your work will become obvious to you. These are fundamental elements of car accidents, bar fights, gunshots, and almost everything else your day on the job may bring.

I've been asked by attorneys to estimate the time a photograph was taken (which might implicate the defendant, depending on the answer) based only on the lengths of shadows cast by the Sun within it. For that line of work, you also need an astro course: "Astro 101," which is surely offered right alongside physics, wherever you choose to go.

The act of solving the course's homework problems is, in a way, a slow re-wiring of your brain—ultimately empowering you to use a lens of investigation that's forged in the operations of nature.

If you have not taken physics before, nor the math to go with it, then these classes may be hard. But if doing things "because they're easy" were part of your identity, I don't think you would have chosen to be a cop in the first place.

Good luck. In the end, you will not regret a moment of it.

Best to you,
Neil deGrasse Tyson

Gifted Students

In October 2004 I visited Kent State University's Stark campus to participate in their Featured Speakers Series. In my talk I emphasized the value of hard work and ambition to success in school, job, and life. During the Q&A session, Bronwen, a student, asked about the importance of gifted education for smart kids from grades K through 12. In a letter sent a week later, she continued her commentary, noting that she, herself, was identified as "gifted" from the very beginning of elementary school and was persistently ignored by teachers, knowing she'd get an "A" without extra help from them. Because of these forces, she worried that untold numbers of gifted students would never rise to their potential. This formed an occasion for me to expand on my views as well.

Hello Bronwen,
Thank you for your summary reflections.

I have several reactions to your comments.

1. To be gifted in a class of non-gifted students is indeed a recipe for being ignored. But to be gifted in a class of gifted students is, in every case I know (and have heard of) a means for the city/county/state to devote extra resources to the "needs" of the gifted. My comments referred specifically to gifted programs and gifted schools, of which there are many.

2. In my experience, a primary means by which children are identified as academically gifted is by their performance on IQ tests, standardized exams, and other administered tests. If school is to prepare us to achieve in society, then, above a certain minimum level of performance on these means of assessment, your scores are irrelevant to the kind of citizen you become—in adulthood and in the professional workplace, after your first job, nobody asks or cares about your GPA or IQ or SAT scores. I invite you to walk up to any older person (30+ years of age) and pose that question.

3. Let's go on a reasonable, but yet-to-be-confirmed assumption that ambition, as expressed through adulthood, does not directly correlate with GPA—like I said, if

it did, then all of society's greatest achievers (entrepreneurs, lawyers, actors, comedians, artists, athletes, architects, musicians, statesmen, generals, CEOs, presidents, mayors, senators, governors, community leaders, authors, directors, producers, etc.) would be occupied only (or primarily) by those who got straight A's in school. But this is simply not the case. So if ambition can be found in almost any package, and since by design, there are more people who do not get A's than those who do, then perhaps somebody (a dedicated educator) ought to be in search of those who have it. Or better yet, design curricula that develops it.

4. Not that you have any obligation to follow my advice, but in my opinion, if you want to make your mark as an educator, then why not explore how to assess, or otherwise code for ambition, and nurture those students? This will be of incomparably greater value to society than you chasing after "smart" kids just because they are smart.

As a minimum, all I ask is that the label "gifted" be changed to "kids who work hard," so that the club does not come across as exclusive and impenetrable to those outside of it.

All the best in your school and career.
Neil deGrasse Tyson

Accuracy

Saturday, September 25, 2004
Via Email, Forwarded from the
Natural History *Magazine Inbox*

Dear Sir/Madam:
I am writing to you from the U.S. Academic Decathlon (USAD). Last year, you granted our organization permission to publish an article titled "Dust to Dust," by Neil deGrasse Tyson, published in the May 2003 edition of Natural History *magazine, in our curriculum materials.*

We have since received a couple of complaints from one of our coaches about the accuracy of the content. I am very much hoping that the information in the article is correct, and the coach is incorrect, as I would hate to have to issue any corrections.

Here is one of several examples from the Curriculum Director:

> *Your article says that the Sun will eventually become a red giant and swell "a hundredfold in size." One of our coaches feels this is incorrect: When the sun becomes a red giant it will swell to the present orbit of the Earth, 93 million miles from the sun. The sun's present diameter is 864,000 miles. If it increases in size a hundredfold, it will be 86 million miles in diameter. That means its radius will be 43 million miles, less than half the distance to the Earth's orbit.*

Would it be possible for someone to look into his concerns. Thank you in advance for your time and assistance. I look forward to hearing from you.

Sincerely,
Terry McKiernan

Dear Mr. McKiernan,
Thank you for your enquiry. You raise important questions that relate not only to the accuracy of quantities reported in my essay but also to the accuracy of astrophysical quantities in general.

Astrophysics is unique among the sciences for the sheer range of numerical values represented in the objects and phenomena we quantify. For example, the ages of stars span from hundreds of thousands of years to hundreds of trillions of years, depending primarily on the mass but also on other factors.

The temperatures of stars range from a thousand degrees on the surface of the "coolest" stars, to nearly a billion degrees in the cores of the hottest stars.

The wavelength of the longest measured radio waves are meters across, yet the wavelength of the shortest gamma rays are less than one hundred billionth of a meter.

Most things we measure or quantify in everyday life do not span this breadth. So if you get half-off on your store purchase, or if one item is twice as large as another, or if one object moves three times as fast as the other, or contains half as many items, we psychologically think these represent large differences. In astrophysics, however, these differences are small, knowing that the measured properties of things can range by *factors* of hundreds, thousands, or even billions.

When communicating with each other in astrophysics, we invoke high precision only if some other physical quantity depends upon it. Otherwise, the precision is not only distracting, but in most cases, observationally or theoretically unjustified.

When the Sun dies, in about five billion years, it will swell to be so large that it will engulf the inner planets. The "edge" of such bulbous entities are actually poorly defined—where is the edge of the cirrus cloud overhead? Where is the edge of the fog through which you drive? The limit of Earth's atmosphere has no sharp boundary either, so people pick a value that suits their needs. That's why, if you look up the size of Earth's atmosphere in multiple (independent) places, you will likely find very different answers, none of which are wrong.

In another example, a question as simple as, "how many planets are there in the solar system?" does not have an unambiguous answer. Six moons, including our Moon, are bigger than Pluto. Not only that, several objects in the outer solar system are almost the same size of Pluto (within a factor of two in size). So what matters more than "how many?" is "what are their various properties?" and "what features do they have in common?"

And how about the question of when Isaac Newton was born? This too, does not have an unambiguous answer. According to his mother and all local records, he was born on December 25, 1642. But at the time, England (where Newton was born) used the Julian calendar. Today we use the Gregorian calendar (introduced by Pope Gregory in 1582), which is shifted by ten days from the Julian calendar, and not yet adopted in Protestant England in Newton's time. The ten-day difference would place Newton's birthday on January 4, 1643, on the Gregorian calendar. The two answers are different and legitimate—he was indeed born on Christmas day in England.

All this leads to my final point. In spite of how science is taught in grade-school, or in spite of what the public thinks, science is not so much about getting the right answer as it is about getting the right idea. In a contrived but illustrative example, if asked to spell "cat" on a spelling bee, and you answered "k-a-t," you would, of course, be marked wrong even though k-a-t is the actual phonetic spelling of the word. The problem here is that you would be marked equally as wrong had you spelled it "z-w-q." I consider this fact a shortcoming of our educational system, where we are not trained how to think but what to know.

So perhaps for future competitions, when science is the subject, you should find questions that test understanding rather than numerical precision. You would be doing a service to the next generation of students as well as to the intellectual capital of this Nation.

Sincerely,
Neil deGrasse Tyson

Chapter 11

Parenting

Newly born children do not arrive with instruction manuals. And while hundreds of professions require that you become pre-certified, a new parent, with no experience at all, is expected to raise a healthy, productive child by what amounts to on-the-job training. This fact magnifies the value of shared wisdom among parents, all trying to do the best job they can. At times, the challenges to success can seem endless.

Doing Time

Sunday, May 15, 2016
Communication via US Postal Service

Dear Neil deGrasse Tyson
As a father of two bright teens, I write seeking your advice on how to encourage their STEM studies.

I am serving a 92-month sentence at San Quentin for gross negligence vehicular manslaughter, with an anticipated release date in late 2019. Accordingly, I have very limited opportunities to communicate with my precious children—no internet, phone calls limited to 15 minutes, and intermittent visits. I want to encourage my children to study science and math. Given their keen interest in astronomy (one seeks to become the "first astronaut veterinarian"), I hope you could direct me to resources, websites, or organizations my children may use to grow and learn, given their potential.

The consequences of my crime are manifold and impact my children in more ways than I can fully appreciate from San Quentin. Still, I hope to stay involved in their continued development. Any recommendations you could make would be greatly appreciated.

Might my kids visit you in New York? A chance to see a famous scientist, arranged by their father, would be evidence of my continued love for my children—and a special adventure as well.

Regards
Wayne Boatwright, CDC No. AN0094
San Quentin, California

Communication via US Postal Service

Dear Mr. Boatwright,
One of the great revelations of parenting: when you have curious, motivated children, the intervention of a grownup carries almost as much risk in squashing their ambitions as it does in nurturing them. Deep down we know it's true. As the saying goes, we spend the first years of children's lives teaching them to talk and walk, and the rest of their lives telling them to shut up and sit down.

Also, to all our collective dismay, research persistently shows that parents have only marginal effect on the personality that their kids ultimately develop.

At your kids' ages they're surely internet savvy. NASA is not a hidden entity in the media universe, and YouTube abounds with clever and interesting science videos. So I have no doubt your kids are plugged into the moving frontier of science, in proportion to their depths of curiosity.

As for becoming an Astronaut Veterinarian, I don't know how soon we will be taking pets or farm animals into space. But when that day comes, space will have become a routine destination, and we'll probably need tons of Space Vets.

Rather than arrange to have your kids visit on their next trip to NYC, let's wait until you get out, and perhaps you can bring them yourself. You then become part of your kid's memory of that visit.

And if such a trip is not in the near future, I give public talks in California with some frequency, with San Francisco as one of my most loyal fanbases. I will be delighted to meet and greet your two kids when such an occasion arises.

Until then, as always, keep looking up.
Neil deGrasse Tyson

Postscript: Working hard to rehabilitate himself, Wayne Boatwright would be released 500 days early having earned "good time" credits, and has since started a Facebook group, The San Quentin News Crew, to serve as a model for his prison peers.

On Pretending

Monday, March 23, 2009
Dear Neil,
I want my boy to be like you, so I will have to pretend I don't like you.

Thank you for representing being smart in such a favorable way.

A weak astronomy student,
Doug Fedinick

Dear Doug,
Whatever it takes.

Neil deGrasse Tyson

Starry Starry Night

Tuesday, March 24, 2009
Dear Neil,
When I was a kid, my father and I would sit on top of the big green family station wagon and look at the night sky. We would find the constellations and I would make up my own. My favorite was the Fat Hobbit. I haven't stopped looking. Now my father is coming to live with me. I don't have a station wagon, but I do have a fabulous telescope, which I use to get other people looking up. And when my dad gets here, once again we're going to go outside, just the two of us, and look at the night sky.
Lizdel Collado

Dear Lizdel,
Thank you for sharing your personal and touching reflections.

All the best, under the
canopy of your starry skies.
Neil

Home Schooled

Many Christian parents who homeschool their children do so to ensure that the curriculum is rich in Biblical views of the natural world. This often throws established science into question, especially the subjects of evolutionary biology and the origins of the universe. Lisa McLean lived in a religious community where she was homeschooling her daughter, and she was torn by what the religious curricula told her compared with what the discoveries of science reveal. In August 2005 she asked how I handle these conflicts with my own kids.

Dear Lisa,
Thank you for your candid letter.

You asked what I teach my children. My answer is—I do not worry about what they know as much as I worry about how they think. This just might be the highest of all pedagogical goals, because the most important moments in life occur at times when how we think will matter more than what we know.

Teaching someone how to think is hard, and takes more effort on the part of the teacher and student. Among other things, it encourages them to ask questions. It involves being comfortable with ignorance, if that happens to be our collective state of knowledge at the time. It involves experiment and inquiry.

I do not teach my kids about magnetism. I just give them a sack of magnets and tell them to go and play.

I do not teach my kids about centrifugal force. I take them to an amusement park and go on the spinning rides with them.

I do not teach my kids about chemistry, I simply ask them, Have you ever mixed baking soda and lemon juice together? (This combination creates quite a chemical reaction—try it with your daughter.)

When their flashlight doesn't work, I do not say, "It needs more batteries," I say, "Let's test the batteries to see if they are dead." We then put the batteries in a battery tester to investigate first-hand.

When they ask me a question that I do not know, I reply, "Let's find out," and we go to a book or surf the Internet for answers.

If they believe something in the absence of evidence, I ask them, "Why do you believe this?" or "How do you know this?"

For example, right now, my daughter is transitioning out of the tooth fairy stage. She now thinks the tooth fairy has been her parents all these years. This prompted a major discussion in her class. So they proposed an experiment to test the idea. The next person whose tooth came out would not notify his or her parents, but would simply take home the tooth and put it under the pillow. Any real tooth fairy would know about this. But parents would not. If there's no money in the morning, then the experiment strongly argues against the existence of a tooth fairy. That is an example of how to think mattering more than what to think.

As for your direct questions—the Big Bang is the most successful theory of cosmic origins ever put forth, and has reached consensus in the astrophysics community. We are now on to other problems. The public perception that scientists go from one truth to another is simply false. In the modern era of science (i.e. the past 400 years)—that is, in the era of experimental science, when a theory has been resoundingly supported by data—that theory does not one day, all of a sudden, become wrong. The worst that can happen to the theory is that it gets incorporated in to a larger, more powerful idea of how the universe works. So the Big Bang is here to stay, in its current form or in a form where it plays a part of a larger cosmic understanding.

By the way, religious documents are generally referred to as "revealed truths." And true believers of them hold those documents to be divine and unerring. This has caused nothing but trouble in the history of human culture, particularly when two different religious groups hold conflicting ideas of what is "truth."

So, in my judgment, the word "truth" will not serve your child's needs as much as the word "investigate," or, better yet, the word "explore."

With best wishes to you and your family,

Neil

Scary Smart

Wednesday, July 22, 2009
Dear Dr. Tyson and any other kind brainiacs,
My Aspergers son Jack is scary smart and could very well be the next Einstein, which is his nickname. I'm trying to reach out to other super smart scientists who may be able to help Jack grow his gift. Jack's vocabulary and obsessions consist of things like concept cars, nuclear fusion, biotech, particle accelerators, dark matter, anti-matter, worm holes, black holes, nanobots, creating cures for disease, and lots of hydrogen! I have no means of nourishing Jack's brain. His spirit has been almost extinguished by his public school environment.*

I want so badly for Jack to have connections with other people. That can't happen when the people surrounding him can't relate or understand, or believe what he is talking about. Jack is almost 15 and on the verge of serious depression due to his struggles, loneliness, and feelings of inadequacy. It makes me very sad to think that he may never get an opportunity to do something BIG for this planet.

Jack's mom

Dear Jack's mom,
People with borderline Aspergers are not uncommon in the physical sciences, a collection of fields (chemistry, physics, engineering, astrophysics, geology, etc.) in which social talents carries less currency than intellectual development.

* Name changed.

Consider further that in almost every example of an academic professional in my field, high grades were the norm. Possibly a third to a half of my department was the valedictorian of their high school. And in nearly all of those cases, their primary intellectual stimulation came not from school, but from books they read at home—alone. That mode of solitude learning was true for me as well. So your urge to get the public school to serve his needs and interests may be futile. And in the absence of a private school option for him, unlimited access to books and full computer access to the Internet may be your best option for him.

As you surely know, one can build an impressive home library with not much money if you regularly peruse the "remainder" tables of book stores—paying anywhere from $1 to $10 per book on all subjects of the world.

Beyond these measures, from what I know, I will not assert that you have an easy task ahead of you. But it's surely not hopeless.

With best wishes,
Neil

Half Black

Monday, March 23, 2009
Dear Dr. Tyson,
I want to bring my children to New York City. I want to kindle within them the thirst for science. To help me meet this goal, I was hoping you might clue me in on the best days

to bring them. I want to encourage, within my children, a love for learning the sciences—not a fear or disdain for all things scientific.

And, since my children are one-half Black, I want them to have you as one of their role models of possibilities. Often, there is much on TV and online that shows them the negative side of their race, and I want to counteract this with positive influences.

So, when do you think would be a good time to bring them to your city, and hopefully spark within their little souls, a passion for science?

Cathy L. Jones

Dear Cathy,

I take the unorthodox view that the concept of a role model is highly overrated. Or rather, role models should be assembled à la carte. I have found skin-color associations, in these modern times, do more harm than good in the maturing child. To pick one person for a role model and not another, for reasons driven by skin color, may preclude entire worlds of ambition from being realized by your kids.

If you come to visit, it should not be because I am labeled as Black. But because I am a scientist/educator, and you care for the science literacy of your children.

Sincerely,
Neil deGrasse Tyson

Bible Stories

Sunday, February 26, 2017

Dear Dr. Tyson.

I wanted to send you a note because I am having a discussion with my 10-year-old son. We have been doing what generations before me have done . . . sending my child to Hebrew School. We send him so we can have him learn about his religion and where he comes from. However, my son, who by the way is on the Autism Spectrum, said to me last night that Hebrew School is ridiculous because he does not believe in G-d, he believes in science. He believes that bible stories just cannot be true. And the truth is, I can't deny that he just might be absolutely right.

When I asked where he got a lot of his ideas, he said "Cosmos," so I know he believes and respects what you teach. (and, I thank you for that!) MY question is . . . are both possible? Do you think that there might be a higher power out there, or that science and faith CAN find common ground?

I ask, because I respect my son enough for him to have his beliefs, and I don't want to impose anything on him that cannot be proven to be true. I know you are a busy man, but I am working on being a good parent.

I am so thankful for you taking the time to read this.

Sincerely,

Ingrid

Friday, March 30, 2018—Passover
Dear Ingrid,
An embarrassingly belated reply to your thoughtful email. The universe has been keeping me quite busy lately, but I do get to all my emails—eventually.

Of course in a free country, within limits, you can raise your children how you please, on whatever belief system you choose. For this reason, most people in the world who are religious, practice the religion of their parents. For example, the chances of Christians raising a child who later becomes Muslim, or a Muslim family raising a child who later becomes Jewish are extremely rare. The children will be more likely to grow up believing in no Gods than in the Gods of other religions.

So the urge to raise your son as a devout, practicing Jew, being one yourself, is entirely normal and natural. But of course you have, at most, only 18 years of direct influence on him. Your son will spend more than eighty percent of his life under a different roof than you.

From what I have seen and encountered, Judaism manifests across a huge range of practices—from emboldened Jews who enthusiastically eat bacon to the various sects of Orthodox Jews who, among other practices, maintain separate kitchen utensils for dairy and for meat. As a scientist, I have much more experience with atheist Jews. They do not view the Torah as the word of God. They see it as a book of stories—not to be judged for their truth or falsehood, but as a repository of insights from which wisdom for living one's life can be derived.

Think about it—when we read fairy tales, we are not judging them for whether they are true or not.

Instead, we fold lessons derived from them into our world views. Not only this, atheist Jews will commonly celebrate the high holidays with no less ritual than practicing Jews, right on down to leaving an open seat at the Seder table for Elijah, and making sure the front door is unlocked, so he can just walk right in if he happens to show up.

Why would an atheist Jew do this? The answer is not hard. Rituals and traditions account for some of the strongest binding forces among peoples of the world. Attending Mass on Sundays for Catholics. Prayer five times per day for Muslims. Ancestor worship for the Animist religions. One can participate without judging whether the events that established the ritual have any literal truth at all. The participation creates a sense of community, which has almost always contributed value to civilization. It disrupts civilization only when people require that others share their particular rituals, with threat of force to achieve it.

Being on the spectrum and liking science as he does, your best bet might be to not enforce the literalism of anything religious, but to keep him plugged into the beautiful traditions of the religion, and emphasize the value of ritual as a seed and taproot of community. Often that alone represents the greatest challenge when raising autistic children—getting them to embrace the value of love and compassion for people and for relationships.

Rest assured that you can raise a wholesome, intelligent, law-abiding child without requiring he believe that Moses turned a staff into a snake, or that manna fell from heaven.

Good luck. In my experience, it takes some of that too.

Happy Passover to you both.
Neil

First Telescope

Saturday, July 18, 2009
Dear Professor Tyson
I thought I'd pass this along to you, and that you'd appreciate this story more than most people. If not, apologies.

I realized I was over-telescoped and decided to get rid of my 2003 60mm Meade brand refracting telescope. The town of Tombstone, Arizona, is small, and trying to sell it would probably cost more in advertising than anyone would be willing to offer. So, I put up a note in the post office, "Free to any kid 10–17 with parent." Even with "free," it took five days before I got a call.

This guy called and later came over with his 12-year-old daughter. I showed them how the telescope and control-box worked, and all throughout this, the kid's eyes were as big as headlights. I even threw in a spare copy of H.A. Rey's The Stars* *—the first astronomy book my dad gave me, way back in 1955. The kid's entire face consisted of eyes and a smile.*

I never had kids myself, so today I got a brief glimpse

* H. A. Rey, *The Stars: A New Way to See Them* (Boston: Houghton Mifflin, 2008).

into what I missed out on. And this little girl will get a lot of glimpses into a brand new universe.

Fair trade.
MJ "Morg" Staley

Dear Morg,
There's nothing like the right telescope in the right hands of the right person at the right time for the right price.

Neil

Happy 30th Anniversary

16 August 1982
Calligraphy on Parchment

Dear Dad & Mom,

This month I am to receive my master's degree in
Astronomy;
A major achievement of my life which cannot pass
Without the due acknowledgment of two of the most
Warm, caring, and rational people I know.

Central elements of my personality, character,
wisdom and perspective
Are traceable to each of you.
Throughout my twenty-three year quest for the
cosmos
You have never failed to keep my feet on the earth;

To promote my awareness of the aged,
The crippled, the blind, and the other
Inequalities of life and of society.
All the while, your unfatiguing tolerance
Of my interests has found you
Driving many miles for "that particular lens"
Or assisting the transport of my telescopes
In and out of cars, to and from the fields and up and down stairs.

My life has taken me many places;
From twenty-two stories over the Bronx to a snow-carved circle on Peacock Farm,[★]
From the plains of Mojave Desert[†] to the summit of Mount Locke,[‡]
From the Bronx High School of Science to the Harvard College Observatory,
And from Bell Telephone Laboratories[§] to the University of Texas at Austin.
Let there be no doubt that I continually felt your guidance ahead of me, your support behind me, and your love beside me.
For the next thirty years[Δ] may you share each other
The way you have shared yourselves with me.
Happy Anniversary.

—Neil—

★ For one academic year, while I was in seventh grade, my family lived not in a New York City apartment but in a private sublet home in Lexington, Massachusetts, while my father was a one-year fellow at the Kennedy School of Government. We lived on Peacock Farm Road. On one occasion, after a heavy winter storm, I shoveled a backyard path to create a cul-de-sac in the snow large enough to use my first telescope.

† During the summer between my eighth and ninth grades, I attended an astronomy camp for geeky middle and high schoolers, located in the Mojave Desert, southern California, where we lived nocturnally as we observed the clear night skies from a battery of telescopes on location.

‡ The University of Texas at Austin owns and operates the McDonald Observatory in West Texas, which sits atop Mount Locke. I penned this tribute in graduate school, while observing there during the summer of 1982.

§ During the summer between my junior and senior years in college, I was a research intern in the Material Science division at Bell Telephone Laboratories, Murray Hill, New Jersey.

Δ My parents would remain married another thirty-four years, before my father's death at age eighty-eight.

Chapter 12

Rebuttals

Every now and then you just have to fight back.

Making the Grade

My daughter attended and graduated from my high school alma mater. While a junior there, beginning in the fall of 2012, she wanted to take AP calculus, but had not yet taken a formal class in pre-calculus, considered by the Math Department to be a strict pre-requisite. The principal's letter to me was a strongly worded defense of the school's educational mission complete with placement exams, assuring that students never step anywhere academically over their heads, leading to high grades for college admissions.

Who in their right mind would argue with those goals? I did.

To: Principal of the Bronx High School of Science
Thanks for your note, which includes the statement:

"It is our job to help protect
your daughter's GPA"

That's noble. But my life experience tells me it's not the noblest goal out there. As a scientist, an educator, and as a parent, I offer a countervailing quote:

"It's my job to protect her interest in learning"

A true love of learning never ends. Meanwhile, a GPA ends after college and couldn't be more irrelevant to the rest of one's life.

My daughter enjoys math, and wants to take Calculus now, bypassing a year of pre-calculus. She has studied pre-calculus on her own over the summer in anticipation of this. Yet you have a system in place that efficiently prevents this step.

I don't know when it became common for a school to stridently prevent a student from skipping to an advanced class. Especially in an era where promoting interest in STEM fields for girls is a national priority. Most students in most schools take the easiest classes they can find—driven, of course, by the urge to protect their GPA. I'm compelled to note that my mediocre high school GPA prompted no comment from any teacher about whether I would "go far." Yet I had a love of learning that apparently had value to me and to the college of my choice, if not to my high school teachers.

None of us knows for sure how she will do on your pre-calculus placement exam. But dare I recommend that the test score be used as guide, in anticipation of her forthcoming workload in calculus, rather than as a fence, to be kept closed for scores below what you have judged to be acceptable.

If you are worried about my daughter's GPA, don't be. If you are worried about what college she will choose, or what college will choose her, don't be. Worry about what kind of adult she will become. Because what we worry about is whether her High School nurtures a learning environment, without rules designed to impede urges to "go far."

If my daughter does not score well on the placement exam your role should not be to prevent her from advancing but, perhaps, to strongly advise against it. If she wants to proceed in spite of the caution, then your role should be to support the ambition. But even if you can't, or if it's against your pedagogical philosophy, it's not every child who has two parents fluent in calculus,* so her progress in this regard should be the least of your worries as Principal.

Sincerely,
Neil

Postscript: My daughter took the pre-calculus placement exam, offered by the school as a compromise. We do not know how well or poorly she did on it, but they allowed her to take calculus that year, formally skipping pre-calculus. She would score a 5 (the highest grade) on the year-end AP exam.

* My wife has a Ph.D. in mathematical physics.

B.o.B. and the Flat Earth

The popular hip-hop artist B.o.B. ("bee oh bee") is vocal about his flat-Earth beliefs. Early in 2016 he took to social media with his ideas. Normally I would ignore such a foray, but he got my attention when he claimed that the laws of math and physics show that Earth is flat. In the geekiverse, those are fighting words.

Friday, January 27, 2016
Video Letter to the Rapper B.o.B.
Delivered on *The Nightly Show with Larry Wilmore*, Comedy Central

Listen, B.o.B., once and for all, Earth looks flat because, (1) you're not far enough away at your size, and (2) your size isn't large enough relative to Earth to notice any curvature at all. It's a fundamental fact of calculus and non-Euclidean geometry that small sections of large curved surfaces will always look flat to little creatures that crawl upon it.

But this whole thing is just the symptom of a larger problem. There is a growing anti-intellectual strain in this country that may be the beginning of the end of our informed democracy. Of course, in a free society you can and should think whatever you want. And if you want to think the world is flat go right ahead. But if you think the world is flat and you have influence over others, then being wrong becomes being harmful to the health, the wealth, and the security of our citizenry.

Discovery and exploration got us out of the caves and each generation benefits from what previous generations have learned. Isaac Newton said, "If I have seen farther than others it is by standing on the shoulders of giants." So that's right, B.o.B., when you stand on the shoulders of those who came before, you might just see far enough to realize the earth isn't f%@#king flat.

And by the way, this is called gravity. . . .

<< Mic Drop >>

A Horse's Astrophysicist

Unprompted by any major incident, a conservative radio show host and journalist-blogger for a local paper serving Idaho Falls posted an article in August 2016, playfully critical of all I do, with the title "Neil deGrasse Tyson Is a Horse's Astrophysicist." It was filled with political jabs common when people from opposite sides of the political spectrum meet in the boxing ring of social media. His tract labeled me a liberal atheist, and questioned my academic standing as well as my statements on climate change. He also criticized a tweet of mine that referenced the national medal count for the ongoing Olympics, in which I declared that, per capita, some smaller countries were kicking our —. He viewed this and many other liberal tendencies as anti-American. His name is Neal Larson, and I was the first reply in the comment section.

Tuesday, August 23, 2016
Comment Thread

Hello Neal,
First, I'll forgive you for not spelling your name correctly.

Second, and more importantly, I don't mind being labeled a horse's astrophysicist (I see what you did there), provided it's based on factual information. So what we must do is subtract the false information from your article, and then re-assess what name you choose to call me. If it still justifies "horse's astrophysicist," then so be it.

1. FYI: The cosmic perspective is all about having a second look at what, on the surface, appears to be important, special, ego-boosting, etc. The Olympic medal count is not immune from this analysis. In fact an even better measure is medals per GDP per capita. This will tell you how efficiently and effectively a country is spending its wealth on athletic excellence. The Tweet, though based only on population, was a playful plea that we could be winning even more medals than we are.

2. I am agnostic, and actively disavow the atheist label, as multiple on-line videos can attest. Even one rising through 3-million views.

3. People who deny human-induced climate change are badly misinformed. This position is neither politically Liberal nor Conservative. It's factual. Although one could argue that all those who want to preserve the environment are the real conservatives in this discussion.

4. You use "Liberal" as a tag to characterize my politics. Since I have no active public political position, that's a hard task to accomplish. Climate change deniers are misinformed. But so are people who think vaccines gives you autism. And so are people who think genetically modified foods are bad for you. These science-denying postures cross political boundaries.

5. I am 3x appointed by President George W. Bush, serving on commissions to advise him on the future of the American aerospace industry, on NASA, and on the annual Presidential Medal of Science winners. So your disapproval of my views is not shared by others on the conservative spectrum.

6. Lastly: my research output as a scientist is not hidden. You can find an easy link listing all publications on my webpage.

So when you factor in (or simply subtract) all these elements from your writeup, if what remains still justifies labeling me a horse's astrophysicist, then, as I said, I'm okay with that.

Respectfully Submitted,
Neil deGrasse Tyson, New York City

Postscript: Neal Larson was ultimately apologetic and contrite after my reply, both publicly and privately, and we've become email buddies since then. He hosts The Neal Larson Show *on radio and podcast and continues to write occasional newspaper columns.*

Don't Have a Cow

After I posted this short tweet, on Sunday, August 7, 2017:

A cow is a biological machine invented by humans to turn grass into steak.

6:38 PM · Aug 7, 2017

26.4K Retweets **87.6K** Likes

the popular musician (and vegan activist) Moby launched a barbed attack from his Instagram account.

Via Instagram

When one of your heroes breaks your heart. Neil deGrasse Tyson, really? You can Tweet that and make light of the unspeakable suffering experienced by the hundreds of billions of animals killed each year by humans? Or the fact that animal agriculture causes 90% of rainforest deforestation and contributes up to 45% of climate change? Or that according to the World Health Organization and the Harvard Medical School, a diet high in animal products leads to heart disease, cancer, and diabetes. For a smart physicist, Neil deGrasse Tyson, you sound like an ignorant sociopath.

Moby

My reply . . .

Friday, August 18, 2017
Facebook Note
Moby vs Tyson

My cow Tweet was intended to expose a blunt reality: A cow is not a mechanical machine. It's a biological machine. A biological machine with one purpose (actually, of course, two purposes if you include it as a source of milk), and that is to eat grass (or, of course other food stocks), grow big, and be slaughtered for food. They are generally not kept as pets. They don't rescue people in trouble. They do not assist the handicapped. And what's remarkable here is that cows don't exist in the wild. They have never existed in the wild. Farmers genetically engineered them ten thousand years ago from now-extinct ox-like Aurochs in the service of civilization.

So the Tweet is 100% truthful and accurate. The intensity of reactions to it tells me that people presumed I was trying to get them to agree with some opinion I carry. But the Tweet is fundamentally opinion-neutral. Curious that only a few people took the opposite reaction to the cow Tweet, of how diabolical we are to do this with animals, and that it should stop.

I noticed something similar when I posted this opinion-neutral tweet after one of the horrific school shootings some years ago:

Neil deGrasse Tyson
@neiltyson

In Walmart, America's largest gun seller, you can buy an assault rifle. But company policy bans pop music with curse words.

3:00 PM · Dec 22, 2012

13.9K Retweets **3.1K** Likes

The reaction that followed was highly illuminating to me. Presuming me to be an opinion-forcing pundit, people angrily interpreted it their own way, passing judgment on my intent. The reactions divided evenly on whether they thought I was defending (or attacking) the free market, the first amendment's protection of free speech, or the second amendment's protection of gun ownership. A smaller percentage of people, perhaps 20%, saw it as written, with reactions such as, "Thanks. I never thought about that inconsistency!"

If anybody cares about my opinion, I note here that in countries founded on freedom, and where there is resistance to government control of its citizenry (such as the USA), it may be easier to engineer solutions to problems than to get a hundred million people to change their behavior. A possible solution, for which there has been great progress, is the laboratory manufacture of meat proteins, where a person can enjoy a steak that never came from any living creature—a topic explored in a highly popular episode of *StarTalk* that I hosted, featuring the one-and-only Temple Grandin and Paul Shapiro, VP of the Humane Society.

So, I don't quite know what to say to people who react explosively in the face of objective truths, attacking the person who delivers the information. But what's clear is that we now live in a world where differences of opinion lead to fights rather than conversations.

Neil deGrasse Tyson, New York City

Postscript: Moby would later apologize, citing his Instagram post as "unnecessarily harsh."

Keep Off deGrasse

In August 2009 I was chided by Nzingha Shabaka for keeping (and using) my French middle name. Her Afrocentric sensibilities would have none of it, seeing colonial names as a source of low self-esteem in the African-American community. I try to pick my battles, as my reply surely indicates.

Dear Ms. Shabaka,
Thanks for sharing your ardent concerns, but I remain convinced of Shakespeare's aphorism:

"A rose by any other name
would smell as sweet."

I think we should all work hard to ensure that substance matters more than labels—that's the society I strive to live in.

Best to you,
Neil deGrasse Tyson

Hollywood Nights

Wednesday, July 22, 1998
New York Times Op-Ed

Now that New York City is relatively safe from muggers, Hollywood has resorted to monsters and meteors to trigger end-of-the-world fears in urban movie-goers. But unlike romantic comedies or action-adventure thrillers, most disaster films pluck the fruits of science for their storylines. Deadly viruses, out-of-control DNA, evil aliens, and killer asteroids are all common themes in recent films.* Unfortunately, a film's scientific literacy hardly ever measures up to its plot.

Am I the only one who cares?

I'm not talking about simple bloopers, such as when a Roman Centurion happens to be wearing a wrist watch. These mistakes are inadvertent. I'm talking about ignorant bloopers, like reversing the sunset to pretend you have filmed a sunrise. These are not time-symmetric events. Are cinematographers too sleepy to wake up before the Sun and get

* Specifically, *Armageddon*, Touchstone Pictures, 1998, and *Contact*, Warner Bros., 1997.

the real footage? And why is it that movie meteors have such good aim? Earth's surface is seventy percent water and over ninety-nine percent uninhabited, yet an incoming meteor decapitates the Chrysler Building in one of this summer's movies.

And why is it that James Cameron took the time to get every imaginable detail correct in his film *Titanic*—from the rivets to state rooms to the dinnerware—yet he got the wrong nighttime sky? Actually, he comes close. What could be the constellation Corona Borealis (the Northern Crown) is shown overhead on that fateful night. But it has the wrong number of stars. But worse than this, the left half of the sky is the mirror reflection of the right half of the sky. So the universe in *Titanic* is not only wrong, it's lazy.

But why? I'd bet the costumes were researched to be precisely the styles of the period. Had someone been on board wearing love beads, bell bottom jeans, and a large afro, you know that viewers would have complained loudly that Cameron had not done his homework. Am I any less justified in my outcries?

My gripes are not just with Hollywood. What about those majestic stars in the ceiling of New York City's Grand Central Terminal? Rather than just admitting that the backwards constellations were a mistake, a sign in the lobby during renovation tells us, "Said to be backwards, [the ceiling is] actually seen from a point of view outside our solar system." But a second error has now been committed in an attempt to cover up the first: no point of view in our galaxy will reverse the constellation patterns of Earth's night sky. As you leave the solar system, and travel among the stars, all that happens to Earth's constellations is that they become scrambled and wholly unrecognizable.

What society needs are scientifically literate reviewers. Why should a critic be limited to saying things like, "The characters stretched credulity" or "the tonal elements clashed with the emotional flavor of the set designs"? Just once I want to hear a critic say, "Flying saucers don't need runway lights" (as was depicted in *Close Encounters of the Third Kind**), or "The Moon phases grew in the wrong direction" (as what happened in *L.A. Story*†), or "An asteroid the size of Texas would have been discovered two hundred years ago" (as was shown in *Armageddon*). Only then might the public begin to appreciate the role that the laws of physics play in everyday life.

If you want to write a book, make a film, or engage in a public art project, and if this work makes reference to the natural world, all I ask is that you call your neighborhood scientist and chat about it. When you seek "scientific license" to distort the laws of nature, I would prefer you did so knowing the truth, rather than inventing a storyline that is cloaked in ignorance. You may be surprised to learn that valid science can make fertile additions to your storytelling—whether or not your artistic objective is to destroy the world.

Neil deGrasse Tyson
New York City

* *Close Encounters of the Third Kind*, Columbia Pictures, 1977.

† *L.A. Story*, TriStar Pictures, 1991.

Epilogue

A Eulogy, of Sorts

Letter to Dad*

Saturday, January 21, 2017
Repass Remarks

Dear Dad,
Thank you for a lifetime of wisdom you've bestowed upon me, drawn from moments, circumstances, and incidences in your life. With your permission, I'll share a few that for me, rise above all others.

I've never forgotten the story of your high school gym teacher, who highlighted your body-type as one that would not make a good runner in the track and field unit of class. Your reaction? "Nobody is going to tell me what I cannot do with my life." You immediately took up running. You also ran in Hitler's Berlin Stadium for the 1946 "GI Olympics." The post-war world was not yet ready for a traditional Olympics, so this special event contested soldier-athletes of the

* Based on a eulogy delivered to friends and family, Holy Trinity Catholic Church, New York City.

various theaters of conflict around the world. And by college you became world class in middle distance races, at one time capturing the fifth fastest time in the world for the 600-yard run. Drawing upon that example for inspiration, I have overcome the most negative societal forces on my life's ambitions.

I've never forgotten the story of your best friend Johnny Johnson, also a track star, competing in a meet against the New York Athletic Club. In the day, they of course admitted only WASPs, so athletic Blacks or Jews instead competed as teammates for the Pioneer Club, founded for that purpose. As Johnny came around the last turn in the quarter mile, he was ahead of the New York Athletic Club runner by several strides when he overheard the fellow's coach audibly yell to his runner, "Catch that nigger!" Johnny's reply to himself was simple and direct, "This is one nigger he ain't gonna catch!" and lengthened his lead to the finish line. What today might be called microaggressions, back then were parlayed into forces of inspiration to excel. From that example, I've used such occasions in my life to excel beyond even the expectations I hold for myself.

You told of immigrant Grandma's work as a seamstress. Grandpa's work as night watchman for the food service company Horn & Hardart. A good thing, because he would occasionally bring home leftover food when the money was tight. Your stories of strife were never hate-filled. Never bitter. Instead, they were hope-filled and inspirational—conveyed with tentative confidence that the arc of social justice will continue to bend towards righteousness. I carry that vision for society's future into every day of my life.

You studied hard in school, and took your interest in social justice all the way to your appointment as Mayor Lindsey's Commissioner of New York City's Human Resources Administration. Journalists don't write articles about news that does not happen. But the programs you enabled in the inner city, empowering the youth during the powder-keg years of the late 1960s, ensured that any unrest or disturbance would be mild. Sure enough, New York was calm compared with what went down in Watts, Newark, Detroit, Cincinnati, Milwaukee, and especially in Chicago, Washington, DC and Baltimore, for which Federal troops were called to quell the violence. You worked behind the scenes on this, with your only reward the quiet knowledge that the nation's largest city did not burn during the most turbulent years of the most turbulent decade in American history since the Civil War. Striving to do what is right, without regard to who takes notice, should be a model for us all.

Your stories and perspectives about navigating the people, politics, funding streams, and legacies of institutions deeply informed my (successful) efforts to create, from whole cloth, a brand new Department of Astrophysics at the American Museum of Natural History. You taught me that in life, it's not good enough to be right. You also must be effective. For that, I now count the formation of that department as one of the highest achievements of my professional career.

So Dad, this "thank you" letter in death is simply

public notice of what I've already thanked you for in life: bestowing upon me guiding principles for living my life to the fullest, and along the way, when possible, lessening the suffering of others.

I know I will miss you, because I already do.

Cyril DeGrasse Tyson
October 1927 – December 2016

Acknowledgments

I thank my literary agent B. Lerner for her support and enthusiasm from the beginning to the end of this project. I also thank L. Mullen, for the persistent bookkeeping necessary to assemble the archive from which this book is drawn, along with my office assistants M. Gambardella and E. Stachow, who served as backup support at every turn. I further thank my editor J. Glusman of W. W. Norton, who continues to value our publishing relationship. I also thank N. Reagan and T. Disotell for their anthropological expertise and S. Soter for his characteristically critical read of the entire manuscript. Most importantly, I thank all those whose letters were included for their permission to reproduce our correspondence. Some of these queries are personal and sensitive, addressing one's fickle and often challenging pathways to happiness and success. Their inclusion in this volume just might benefit others with identical or similar trajectories through life.

Index of Subjects

Aliens

Astronomy

Biology

Buddhism

Careers

Christianity

Conspiracy

Death and Dying

Education

Films

God

Islam

Judaism

Mathematics

NASA

Paranormal

Parenting

Philosophy

Physics

Planets

Politics

Pseudoscience

Race

Religion

World Trade Center

16 August 1982

Dear Dad & Mom,

This month I am to receive
my masters degree in Astronomy;
A major achievement of my life which cannot pass
Without the due acknowledgement of two of the most
Warm, caring, and rational people I know.

Central elements of my personality, character,
wisdom and perspective
Are traceable to each of you.
Throughout my twenty-three year quest for the cosmos
You have never failed to keep my feet on the earth;
To promote my awareness of the aged,
The crippled, the blind, and the other
Inequities of life and of society.
All the while, your unfatiguing tolerance
Of my interests has found you
Driving many miles for "that particular lens"
Or assisting the transport of my telescopes
In and out of cars, to and from the fields
and up and down stairs.

My life has taken me many places;
From twenty-two stories over the Bronx to
a snow-carved circle in Peacock Farm,
From the plains of Mojave Desert to
the summit of Mount Locke,
From the Bronx High School of Science to
the Harvard College Observatory,
And from Bell Telephone Laboratories to
the University of Texas at Austin.
Let there be no doubt that I continually felt
your guidance ahead of me,
your support behind me and
your love beside me.
For the next thirty years may you share eachother
The way you have shared yourselves with me.
Happy Anniversary.

~Neil~